THE PSYCHIATRIC STUDY OF JESUS

The Psychiatric Study of Jesus

Exposition and Criticism

By

ALBERT SCHWEITZER

Translation and Introduction by
CHARLES R. JOY

Foreword by
WINFRED OVERHOLSER

THE BEACON PRESS · BOSTON

In the knowledge that he is the coming son of man, Jesus lays hold of the wheel of the world to set it moving on that last revolution which is to bring all ordinary history to a close. It refuses to turn, and he throws himself upon it. Then it does turn and crushes him. Instead of bringing in the eschatological conditions, he has destroyed them. The wheel rolls onward, and the mangled body of the one immeasurably great man who was strong enough to think of himself as the spiritual ruler of mankind and to bend history to his purpose, is hanging upon it still. That is his victory and his reign.

—Albert Schweitzer

PREFATORY NOTE

Albert Schweitzer passed his state medical examinations in the fall of 1911. The surgeon who held the final examination, Madelung, told him that it was only because of his excellent health that he had gotten through a job like that. The previous month he had played the organ accompaniment for Widor's new *Symphonia Sacra*, Widor himself leading the orchestra. Schweitzer's fee for his organ work was used to pay for the medical examination.

There was still his internship to complete and his doctoral thesis to submit before his degree was assured. For the latter he chose to write on recent medical treatises which showed, at least to the authors' satisfaction, that Jesus was mentally diseased. While Schweitzer's book was mainly concerned with the work of three different men he felt it his duty to familiarize himself with the whole literature of paranoia before undertaking his task. The result was that this small treatise took him over a year to complete. On several occasions he was on the point of choosing an easier subject for his thesis.

The book was published in German in 1913. It was first translated into English by W. Montgomery under the title "The Sanity of the Eschatological Jesus," and published in the British periodical, *The Expositor*, VIII series, vol. 6. The present volume is a fresh translation, and constitutes the first appearance of this important work in book form. For assistance in the translation of technical psychiatric phraseology, the translator wishes to record his debt to Dr. Winfred Overholser.

C.R.J.

Newton Highlands
Massachusetts

CONTENTS

	PAGE
PREFATORY NOTE	7
FOREWORD	11
INTRODUCTION	
Schweitzer's Conception of Jesus	17
PREFACE TO THE 1913 EDITION	27
THE PSYCHIATRIC STUDY OF JESUS	33
INDEX	75

FOREWORD

The 19th century gave birth to many scientific advances, to applications of the scientific method of inquiry in various fields. The readiness to demand evidence which is an essential part of the scientific process, and which had already upset man's ideas of geocentricity (Copernicus), ecclesiastical authority (Luther), and original creation (Darwin), began to be applied to the study of history. To this study of history the Bible was not immune, and late in the 17th century (1670) there came about the beginnings of the "higher criticism," with the appearance of Spinoza's *Tractatus Theologico-Politicus*. Some of this "higher criticism" was basically hostile to established belief, but the better part of it was healthily skeptical instead; there was a genuine seeking for historic truth and for a study of motives.

It was inevitable that in the quest for motives some consideration should be given to the possibility that the beliefs of Jesus might be explained as those of a mentally abnormal person, perhaps even of one clearly deranged. Possibly the merely nascent state of psychiatry furnished one reason why more of the iconoclasts did not venture earlier on this path of inquiry. Noack (*Die Geschichte Jesu*, 2nd ed., 1876) referred to Jesus as an "ecstatic," but did not impute mental disease to him—that was left for the 20th century.

In the first two decades of the present century no less

than three medical writers embarked upon a psychiatric "interpretation" of Jesus—a German, Dr. Georg Lomer, who wrote under the pseudonym of George de Loosten; a French writer, Charles Binet-Sanglé; and an American, Dr. William Hirsch. A fourth writer, Emil Rasmussen, Ph.D., included Jesus among a group of prophets whom he classified as psychopathological types. It is to a refutation of these four books that our author dedicates this volume, his thesis offered for the degree of Doctor of Medicine at Strassburg University in 1913. Dr. Schweitzer, already the holder of degrees in philosophy and divinity, had shown himself a sound historian in his *Geschichte der Leben-Jesu-Forschung* in 1906; in his present study he marshals his historical data effectively, together with the knowledge of mental disorder as it then existed in Europe.

Since the authors discussed by Dr. Schweitzer agree on one point, namely that Jesus suffered from some form of "paranoia," a few words concerning this type of mental disorder may not be out of place. The word is an old one—it was used in the Hippocratic writings, though in a general sense, as meaning mental disease. It was introduced into German psychiatry as early as 1818 by Heinroth, but with so loose a definition that at one time from 70 to 80 percent of the patients in European mental hospitals were diagnosed as suffering from "paranoia." Indeed, as late as 1887 a French psychiatrist (Séglas) referred to it as a word which had "la signification la plus vaste et la plus mal définie." Gradually it came to include a variety of clinical groups characterized by ideas of persecution and grandeur, in varying proportions. Some of these groups exhibited almost entirely a distortion and misinterpretation of actual facts, others some elaboration with fabrication, while some showed such a loss of contact with reality as to cause the patient to suffer from hallucinations in one or more of the sensory spheres. A religious coloring of the delusions is far from uncommon. Kraepelin, the great German descriptive psychiatrist, defined these various groups—paranoia, paraphrenia (now generally referred to as paranoid condition) and dementia praecox of the paranoid type, his final formulation appearing about 1913. To Kraepelin and his school, as to the French school of psychiatry, paranoia was largely a question of constitution; it was based on the makeup of the person, developed insidiously and progressively, and was essentially unamenable to treatment. They looked on it as almost if not quite entirely a disturbance of the intellectual functions. It was only in 1906 that Bleuler emphasized the importance in the disorder of reaction to life situations, as opposed to a fatalistic interpretation, and it was after the appearance of Schweitzer's answer to the psychiatrists that a more dynamic interpretation of the mechanisms of paranoia and the paranoid conditions came about as a result of Freud's penetrating observations. (Freud's notes on the Schreber case, published in 1911, were very likely unknown to Schweitzer as he wrote.) We now know, of course, that the emotional and homosexual factors are highly important, and that paranoia is no more a purely intellectual disorder than any other psychosis.

Two general comments may be made on psychiatric diagnosis. First, one must have a good case history. In the case of Jesus, we have virtually none. The first Gospels were probably written 40 or more years after the death of Jesus; as such, their accuracy as to detail, at least as psychiatric documents, must be questioned. Furthermore, we know nothing except by tradition of any but the last year or so (at the most) of Jesus' life, and in all accounts there is a gap of at least 18 years. We know very little of his relations with his mother and siblings and, while we know something of the multifarious social, religious and economic influences of the time, we know very little of the manner in which they played upon him and moulded his feelings and reactions. The perils of diagnosis *à distance* are great!

In the second place, we cannot study any patient in a vacuum, for nobody lives in one, and everyone is conditioned by his environment—its religious and social ways and beliefs, *inter alia*. Schweitzer discusses very effectively the religious beliefs of Jesus' time as large factors in his ministry—the present control of the world by evil spirits, the coming of the Messiah and his Kingdom, the Messiah's suffering, the judgment, the resurrection, and the transfiguration of nature—which were familiar to Jesus and his hearers, and which were as important in the life of the community as our belief in democracy and the Constitution, for example. To the Haitian native a belief that necromancy may be employed against him is a part of the folkways, is "normal"—for the educated resident of Park Avenue such a belief would be properly classified as delusion. In the same way to take Jesus' beliefs and teachings out of their context, as Binet-Sanglé and the other writers we have mentioned do, is psychiatrically unsound.

Schweitzer comments upon the fact that hallucinations —that is, sensory impressions without an external source— are not to be found only in the mentally ill. His statement is entirely correct. Such experiences are not uncommon in a state of consciousness between sleeping and awaking; they are referred to as hypnagogic hallucinations. Some persons develop hallucinations with relatively small doses of therapeutic drugs, such as atropine. Again, in persons of strongly religious natures hallucinations may be found in periods of intense religious emotion, as pointed out for example, by William James. The recorded hallucinations of Jesus, even assuming their historicity, then, are far from establishing the thesis of Binet-Sanglé, Hirsch and de Loosten.

One may disagree with Schweitzer on one or two minor points. He takes for granted that the failure of Jesus to develop ideas of injury and persecution rules out the possibility of a paranoid psychosis. This is not necessarily true; some paranoids manifest ideas of grandeur almost entirely, and we find patients whose grandeur is very largely of a religious nature, such as their belief that they are directly instructed by God to convert the world or perform miracles. Again, he offers as evidence of freedom from paranoia the fact that Jesus modifies his views as to his missions. Some paranoids substantially modify their delusions in accordance with their view of environmental factors, and may indeed appear to reason logically concerning events of interest to them—logically, that is, if one grants their premises.

These are, however, far from fundamental points of disagreement. Dr. Schweitzer's presentation exhibits a profundity of scholarship, theological, historical, and medical, and at the same time the deepest possible reverence for the meaning and the message of the Man of Nazareth.

—WINFRED OVERHOLSER, M.D.
President, American Psychiatric Association

Washington
District of Columbia
1948

BIBLIOGRAPHY FOR FOREWORD

Freud, S.: *Certain Neurotic Mechanisms in Jealousy.* "Paranoia and Homosexuality," pp. 232–243. From Collected Papers, Vol. II, London, 1925.

Freud, S.: *Psychoanalytic Notes upon an Autobiography.* "Account of a Case of Paranoia (dementia paranoides)." 1911, pp. 387–470. From Collected Papers, Vol. III, London, 1925.

James, W.: *Varieties of Religious Experience.* New York, 1902.

Jouannais, S.: *Étude Critique des États Paranoides.* (Thesis, Univ. of Paris), Paris, 1940.

Kirchhof, H.: *Klinischer Beitrag zur Differentialdiagnose paranoider Erkrankungen.* (Thesis, Univ. of Berlin), pp. 45, Berlin, 1937.

Kraepelin, E.: *Manic Depressive Insanity and Paranoia.* Edinburgh, E. & S. Livingstone, 1921.

Lacan, J.: *De la Psychose Paranoiaque,* pp. 381, Paris, 1932.

Noyes, A. P.: *Modern Clinical Psychiatry.* Ch. 27. Philadelphia, 1939.

Peixoto, A. et Moreira, J.: *La Paranoia légitime: son origine et nature* (1906), pp. 63–83. In Peixoto, A.: *Paranoia.* São Paulo, Brazil, 1942.

INTRODUCTION

Schweitzer's Conception of Jesus

Albert Schweitzer is one of earth's true noblemen and his nobility is shown in his touching humility. As simple as a child, as humble as the grass, he has gone his chosen way quite oblivious to the antiphonal choruses of praise and blame which have surrounded him. Those who have come to acclaim him almost as a saint today have often forgotten that he has been the center of violent controversy in every sphere where he has moved so quietly, so sincerely, so unobtrusively.

His interpretation of Bach was new and revolutionary although undoubtedly it was a return to the spirit and to the meaning of the cantor of Saint Thomas. That it was not the current view in the musical world is suggested by an unpublished doctoral thesis on the shelves of the Widener Library at Harvard University entitled "The Schweitzerian Heresy." The heresy alluded to is a musical, not a theological, heterodoxy.

As a philosopher of civilization, he has sometimes struck the note, as Dr. A. A. Roback has said, "of a Jeremiah sitting amidst the ruins of his battered city-state." With gentle but incisive clarity he has condemned our modern civilization with its inhuman mechanisms, its herd mentality, its scorn of human values, its thought paralysis, its spiritual bankruptcy. He prophesies its almost certain decay and sounds a note of warning that there are no other civilizations to take the place of this present one if we fail to rescue it.

Yet nowhere else has this simple, quiet, humble man provoked so stirring a controversy, so violent an upheaval, as he has in the realm of theology. His study of the historic personality of Christ has been described by one commentator as "indeed a drama—a tragedy some would say." More than forty years ago his book, *The Quest of the Historical Jesus,* appeared and still the circles widen from the initial disturbance caused by the falling of this book into the placid surface of modern historical theology.

Schweitzer himself was well aware of this. In his book he pointed out that there could be no compromise between the liberal theology of modern times in its treatment of the life of Jesus and the eschatological Jesus, the Jesus who believed in the abrupt coming of the Messianic Kingdom, which he had discovered in the New Testament sources. He said: "There has entered into the domain of the theology of the day a force with which it cannot possibly ally itself. Its whole territory is threatened. It must either reconquer it step by step or else surrender it." "Modern theology," he goes on, "is no doubt still far from recognizing this. It is warned that the dyke is letting in water and sends a couple of masons to repair the leak; as if the leak did not mean that the whole masonry is undermined and must be rebuilt from the foundation."

No atomic bomb could have had so shattering an effect upon a city as this thought of Jesus had upon the strong citadel of modern theology. Indeed, its effect was utterly devastating. The liberal conception of Jesus current at the beginning of the twentieth century disintegrated under the impact of it. It lay in ruins as abject as those of Hiroshima after the death dealing attack from the skies. But the ruins were alive with a new and strange force like the radio-activity in the desolate Japanese city. The liberalized, modernized, unreal, never existing Jesus, the Jesus which we had constructed to harmonize with our own ideals of life and conduct, gave way to the majesty of the heroic figure who can be comprehended only in the light of his time, who shared all the ardent expectations of his day, who shaped his life and his teaching, and even planned his death in accordance with certain Messianic convictions which subsequent events proved to be completely mistaken. Yet there remains something superb, something dynamic about this historic figure. Out of the ruins a greater city is rising, a strong city, a lovely city on which the glowing rays of the rising sun of truth begin to shine.

Schweitzer himself said of his work: "The judgment passed upon this realistic account of the life of Jesus may be very diverse, according to the dogmatic, historical, or literary point of view of the critics. Only with the aim of the book may they not find fault: to depict the figure of Jesus in its overwhelming heroic greatness and to impress it upon the modern age and upon the modern theology." And again: "We must go back to the point where we can feel again the heroic in Jesus. Before that mysterious Person who in the form of his time knew that he was creating upon the foundation of his life and death a moral world which *bears his name*, we must be forced to lay our faces in the dust without daring even to wish to understand his nature. Only then can the heroic in our Christianity and in our 'Weltanschauung' be again revived."

Albert Schweitzer was born in Kaysersberg in Upper Alsace, on January 14, 1875. The war of 1870 had brought Alsace under German rule and as a German citizen the young man was subject to compulsory military service. On April 1, 1894, his year of training began but he was able to continue some of his theological work through an arrangement made with an indulgent captain. In the fall of the year he found himself on maneuvers near Hochfelden in Lower Alsace. It was characteristic of him that the most prized article in his knapsack was a Greek Testament with which he was working in preparation for an examination in the Synoptic Gospels to be taken at the beginning of the winter term. Every evening, every rest day the tireless young man studied. One day in the town of Guggenheim his mind was stirred to intense activity by his reading of the tenth and eleventh chapters of Matthew. As he read and pondered the verses in these chapters his interest and his bewilderment grew.

At that time Heinrich Julius Holtzmann was lecturing about the Synoptic problem on the theological faculty of the University of Strassburg, and his theory that Jesus' life and teaching could best be understood from the gospel of Mark, since Mark was the oldest of the gospels and was the foundation on which Matthew and Luke had built, was generally accepted. But now in these two chapters of Matthew which are not paralleled in Mark, Schweitzer found a situation that could not be harmonized, he thought, with Holtzmann's theories. In Matthew 10, Jesus sends out the disciples to preach throughout the land of Israel, and tells them that they will be sorely persecuted, but that before they have completed their mission the Messianic Kingdom will be ushered in. Jesus, therefore, did not expect to see them again. It turned out, however, that they were not persecuted and that the Messianic Kingdom did not appear before the end of their journey. To Jesus' own amazement they came back to him, his predictions quite unfulfilled.

Holtzmann had said that this discourse was not historical and that it had been put into Jesus' mouth at a later period, but Schweitzer was sure that no later generation would ever have attributed to Jesus statements proved false by events.

INTRODUCTION   21

In Matthew 11, we have the story of the question put to Jesus by John the Baptist, and Jesus' reply. Schweitzer came to believe that when John asks Jesus, "Art thou he that should come?" he was referring, not to the Messiah, but to Elijah, who in popular belief was to precede the Messiah. This question, and Jesus' vague reply to it, prove that neither John, nor the immediate disciples believe him to be the Messiah. After John's messenger has departed, Jesus turns to his disciples and says (Matt. 11:11): "Among them that are born of women there hath not risen a greater than John the Baptist: notwithstanding he that is least in the kingdom of heaven is greater than he." Schweitzer puzzling over these words came to reject the usual interpretation that Jesus was relegating John to a place inferior to the place of the least of his own disciples. His own disciples were born of women. Jesus could not have meant that. No, Jesus was thinking of two different worlds: one, the natural order in which they all then lived, where John was the greatest of all; and one, the Messianic Kingdom that was to come so soon, where the least of the supernatural beings who should people that world would be greater than John.

These conclusions reached in the midst of the military maneuvers changed the whole course of Schweitzer's thinking, and in so doing changed the whole course of modern theology. "When I reached home after the maneuvers entirely new horizons had opened themselves to me. Of this I was certain: that Jesus had announced no kingdom that was to be founded and realized in the natural world by himself and the believers, but one that was to be expected as coming with the almost immediate dawn of a supernatural age."

The key to all the puzzles that Schweitzer had encountered in the gospels he found in Jesus' commissioning of the disciples, in this interchange between John and Jesus, and in the way Jesus acts when the disciples returned from their tour of preaching. For when the disciples return without having suffered any persecution Jesus is quite evidently very much perplexed. The "Woes of the Messiah" have not fallen upon them, the Messianic Kingdom has not appeared. Jesus, then, withdraws with his immediate followers from the multitudes that thronged around him, retires into the north and begins to rethink his position. Now, at last, he comes to believe that he himself must suffer death before the appearance of the Kingdom, and that by his death he will atone for the elect and save them from the days of tribulation. There can be no question that the 53rd chapter of Isaiah with its passage about the servant who suffers for the sins of others has helped to formulate this conviction.

At Caesarea Philippi the new conviction that he is to suffer death and then will become the Messiah when God's Kingdom is ushered in, is disclosed to the disciples, and when he leaves his retirement to join the band of pilgrims from Galilee who go up to Jerusalem, the disciples alone know what he believes himself to be. One of them betrays this secret to the Sanhedrin but Jesus makes his condemnation and death certain by admitting this fact to them when he is brought to trial. They will see him coming on the clouds of heaven and sitting at the right hand of God, the expected Jewish Messiah. This was, indeed, blasphemy. And so, after a public ministry of possibly not more than five or six months Jesus was crucified.

This, in brief outline, is the picture of Jesus which Schweitzer drew as the result of his first discoveries on that rest day in Guggenheim. Gradually he elaborated them, testing them from every possible point of view, attempting to meet every criticism of them, seeking only for the truth in Jesus and his teaching. His studies took the form of books, a brief book on *The Problem of the Last Supper Based upon the Scientific Research of the 19th Century and the Historical Accounts*, published in 1901; a slightly longer book on *The Secret of the Messiahship and the Passion. A Sketch of the Life of Jesus*, also published in 1901; and a much longer book on *The Quest of the Historical Jesus. A Critical Study of Its Progress from Reimarus to Wrede*, published in 1906.

With the publication of these books the old liberal Protestant view of a Jesus who spoke about a Kingdom of Heaven that was to be achieved gradually here on this earth became forever untenable. We all remember that portrait of a Jesus who no longer was the second person of the Holy Trinity but a human being like the rest of us, a human being, however, who saw clearly through the follies and foibles of his time, who was brought up with all the strange ideas of a supernatural Messianic Kingdom, but who was himself emancipated from them. To the liberal Protestant of the late nineteenth and early twentieth century Jesus was the profoundly intelligent, wholly lovable exemplar of human possibilities, "the Lord and Master of us all." We sang, "O thou Great Friend to all the sons of men, who once appeared in humblest guise below, sin to rebuke, to break the captive's chain, and call thy brethren forth from want and woe." It was a pleasant inspiring picture. The only trouble with it was that it was wholly a fiction of the modern imagination. There was no Jesus of that kind pictured in the two most important sources, Matthew and Mark. The Jesus of modern Protestant liberalism was a convenience created to conform.

The Jesus of the New Testament's most trustworthy sources was a Jesus who expected the imminent end of the world, who believed himself to be appointed by God as the ruler of a supernatural Kingdom about to appear. The historical Jesus was not the "gentle Jesus, meek and mild" of our childhood's hymn, but the eschatological Jesus who shared with his time those strange and stirring beliefs on which late Judaism pinned its hopes. Schweitzer believed that most of the riddles about the life and teaching of Jesus could be solved when once we accepted this conception of the man of Nazareth. He knew that this conception would not be popular. He was well aware of its destructive potency. He knew what would happen to conventional Christian piety as a result of it. Yet he was unafraid. "I comforted myself," he said, "with words of St. Paul's which had been familiar to me from childhood: 'We can do nothing against the truth, but for the truth.' Since the essential nature of the spiritual is truth, every new truth means ultimately something won. Truth is under all circumstances more valuable than non-truth, and this must apply to truth in the realm of history as to other kinds of truth. Even if it comes in a guise which piety finds strange and at first makes difficulties for her, the final result can never mean injury; it can only mean greater depth. Religion has, therefore, no reason for trying to avoid coming to terms with historical truth."

One result of his findings Schweitzer perhaps did not anticipate. He could hardly have foreseen that superficial thinkers wearing the garb of psychology and psychiatry would find in his eschatological picture of Jesus support and comfort for their contention that we have to do in this man of Nazareth with mental derangement, with hysteria perhaps, with paranoia certainly. This new school of psychopathology found here a man who suffered from hallucinations, from ideas of reference, from delusions of grandeur. In the first decade of the twentieth century, books appeared which disturbed Schweitzer profoundly. They asked if Jesus was an ecstatic, they frankly pronounced him to be an insane man, they analyzed the Gospels for evidence of psychopathic symptoms. They were very little qualified for their task. They had no critical understanding of the sources. They based their findings on historically discredited material. Yet Schweitzer's friends pointed out to him that his own studies were in part responsible for them. The writing of this little book was, therefore, to Schweitzer an inescapable duty. He himself was sure that Jesus was completely sane. That Jesus shared the Messianic ideas of late Judaism, that he who was really a descendant of David had come to believe that in the world to come he was destined to be the Messiah, are in no rational sense evidences of mental disease. The little book that follows came out of a deep inner compulsion. It had to be written. It made its case.

Schweitzer points out that it is conceivable that religious truth might be preached independently of any age, truth that is universally and everlastingly so, truth that is valid for every succeeding generation and century. The simple fact, however, is that the truth that Jesus taught is not of that kind. He had a gospel of love to teach and then he wrapped it up in the ideas of his contemporaries. We cannot appropriate to ourselves this gospel of love by refusing to recognize the wrapping. Each age must unwrap the gospel and then apply it afresh to itself, which means, in all probability, enveloping it again in temporary covers. After all Jesus is profoundly concerned with love itself without which no man will enter into his Messianic Kingdom. There is, then, this enduring kernel in his teaching which we must learn to make our own in our own way.

Schweitzer himself best expresses it: "Even if liberal Christianity has to give up identifying its belief with the teachings of Jesus in the way it used to think possible, it still has the spirit of Jesus not against it but on its side. Jesus no doubt fits his teaching into the late Jewish Messianic dogma. But he does not think dogmatically. He formulates no doctrine. He is far from judging any man's belief by reference to any standard of dogmatic correctness. Nowhere does he demand of his hearers that they shall sacrifice thinking to believing. Quite the contrary! He bids them meditate upon religion. In the Sermon on the Mount he lets ethics, as the essence of religion, flood their hearts, leading them to judge the value of piety by what it makes of a man from the ethical point of view. Within the Messianic hopes which his hearers carry in their hearts, he kindles the fire of an ethical faith. Thus the Sermon on the Mount becomes the incontestable charter of liberal Christianity. The truth that the ethical is the essence of religion is firmly established on the authority of Jesus.

"Further than this, the religion of love taught by Jesus has been freed from any dogmatism which clung to it by the disappearance of the late Jewish eschatological worldview. The mould in which the casting was made has been broken. We are now at liberty to let the religion of Jesus become a living force in our thought as its purely spiritual and ethical nature demands. We know how much that is precious exists within the ecclesiastical Christianity which has been handed down in Greek dogmas and kept alive by the piety of so many centuries and we hold fast to the Church with love, and reverence and thankfulness. But we belong to her as men who appeal to the saying of St. Paul: 'Where the Spirit of the Lord is, there is liberty,' and who believe that they serve Christianity better by the strength of their devotion to Jesus' religion of love than by acquiescence in all the articles of belief. If the Church has the spirit of Jesus there is room in her for every form of Christian piety, even for that which claims unrestricted liberty.

"I find it no light task to follow my vocation, to put pressure on the Christian faith to reconcile itself in all sincerity with historical truth. But I have devoted myself to it with joy, because I am certain that truthfulness in all things belongs to the spirit of Jesus."

— CHARLES R. JOY

PREFACE TO THE 1913 EDITION

I have undertaken in the present book to examine thoroughly the conjecture which first appeared in David Friedrich Strauss, and which more recently has been repeated by many a historian and doctor, that the Jesus who lived in the world of ideas contained in the Book of Daniel and in the late Jewish apocalyptic literature and who considered himself the "Son of Man" and the "Messiah" soon to appear in supernatural glory, is to be adjudged in some fashion as psychopathic.

I have felt a certain compulsion to undertake this task, since in my *Geschichte der Leben-Jesus-Forschung*[1] (Tübingen, 1906; second edition, 1913), I had brought out the apocalyptic and what in modern concepts is considered the visionary in the Nazarene's thought world more vividly than any of the investigators who formerly worked in this field, and so had been constantly reminded by H. J. Holtzmann and others that I had portrayed a Jesus whose object world looked like a structure of fantasies. There were warning allusions—on occasion—to the medical books which believed that the "paranoia" of the Jewish Messiah had been proved.

---

[1] Translated by W. Montgomery under the title *The Quest of the Historical Jesus* (literally, *History of the Study of the Life of Jesus*). (London, 1910: A. & C. Black, Ltd. 2nd Ed. 1911; reprinted 1922, 1926, 1931. Published in the United States by The Macmillan Company, New York, 1948.)

As I had carried on studies in both historical theology and in medicine, I considered myself qualified to examine the surmises, the judgments, and the books about Jesus' mental condition just as it is possible for me to form an opinion about both the results of the critical and historical investigation and the standards of psychiatry that are applicable.

That I command the impartiality necessary for this undertaking I believe I have proved by my former studies in the field of the life of Jesus. Should it really turn out that Jesus' object world must be considered by the doctor as in some degree the world of a sick man, still this conclusion, regardless of the consequences that follow from it and the shock to many that would result from it must not remain unuttered, since reverence for truth must be exalted above everything else. With this conviction I began the work, suppressing the unpleasant feeling of having to subject a great personality to psychiatric examination, and pondering the truth that what is great and profound in the ethical teachings of Jesus would retain its significance even if the conceptions in his world outlook and some of his actions had to be called more or less diseased.

In the course of my investigation, however, it has become clear to me, that the students who have found something psychopathic about the apocalyptically-minded Jesus have identified with the morbid whatever in his thought world is peculiar and strange to us without questioning more closely whether this apparently obvious identification is either generally or particularly valid in the present case. If one goes into this decisive problem, it develops that the standards established by modern psychiatry do not permit such an identification as the above, which appears in the utterances of historians and doctors concerned with this subject. The completion of the study, which with the exception of additional observations concerning the work of Rasmussen and its broader consequences constitutes the contents of my medical dissertation, would have been impossible without the valuable supervision of Privy Counsellor Professor Doctor Wollenberg (Strassburg) and the farseeing counsel of Professor Doctor Pfersdorff (Strassburg). To both of them I express my heartfelt thanks.

I owe a debt of gratitude also to Pastor Leyrer (Schirmeck, Lower Alsace) and to Licentiate Menegoz (Strassburg) for relieving me of the burden of proofreading.

March, 1913

ALBERT SCHWEITZER

THE PSYCHIATRIC STUDY OF JESUS

*T*HE PSYCHOPATHOLOGICAL METHOD, which conceives its task to be the investigation of the mental aberrations of significant personalities in relation to their works, has recently fallen into disrepute. This is not because of the method, which with proper limitations and in the hands of professional investigators can produce and has produced valuable results, but because it has been faultily pursued by amateurs. The prerequisites which are essential for successful work in this field—exact source knowledge, adequate medical, and particularly psychiatric experience, both under the discipline of critical talents—are very seldom found together.

One often encounters in this field of study, therefore, misconceptions of the grossest kind, caused by the lack of one or another of these prerequisites and sometimes by the lack of all of them.

To form a judgment about any person on the sole basis of his acts is contrary to all psychiatric practice and has always something suspicious about it. If this is true for the present age, how much more restraint must be exercised when we are dealing with people from a very distant epoch and with imperfect and uncertain traditions! For this reason the constantly recurring instances of historical epileptics, like Mohammed, Julius Caesar, and even Napoleon I himself, seem to us very questionable and legendary. Even more uncertain is the ground on which we tread, when we endeavor to investigate in the light of modern psychiatry the minds of people from a far distant epoch.

Because of this, today's psychiatrists are also disinclined for the most part to use the psychopathological method, partly because they do not consider modern psychiatry so perfected and stabilized that they can find a useful criterion in it for all the acts of mankind, partly because they know that every vital human activity must be understood within the conditions of its own age.

Even though the general considerations hereafter mentioned justify certain prejudices against the psychopathological literature, there are also in the nature of the problem special reasons for disinclination and these latter take on an extraordinary emotional value when it is a question of dealing psychopathologically with the life of Jesus, as has actually been done by various people.

All the more clearly indicated, however, seems to be the necessity of assuring oneself here of the prerequisites enumerated above.

The objective of this book is so conceived, then, that the opinions advanced by the medical writers who have dealt with Jesus should be carefully examined from the psychiatric and critical point of view.

The suspicion that the mind of Jesus might somehow have morbid characteristics found utterance in historical research long before psychiatry became interested in the person of the Nazarene. When David Friedrich Strauss began first of all to work directly with the historical material, he felt obliged to declare that Jesus, as his biography reveals, lived with the quixotic idea that he was destined to appear in the near future in a blaze of supernatural glory, surrounded by angels, on the clouds of heaven, to judge the world as the expected Messiah and to establish the Kingdom which was to follow.

In his first *Life of Jesus*[1] he asserts, in speaking of this matter, that Jesus must be considered from our point of view a fanatic. Immediately, however, he tries to explain that the Nazarene, even though that fanatical idea had gripped him, can be considered, nonetheless, as one in full possession of all his faculties, partly because of the fact that his expectation has its roots in the general conceptions of late Judaism.

When he wrote his second *Life of Jesus* in 1864, he was so vividly conscious of the fanatical in the thought of the second coming, that, as he says in a letter to Wilhelm Lang, he was inclined to consider the idea as very close to madness, and accordingly doubted whether the sayings that refer to this really originated with Jesus. Therefore, he decided to let them fall completely into the background in his portrayal of Jesus, and was reproached by various critics for apostasy from the better judgment he showed in 1835.[2]

In the last decade historical research has more and more clearly perceived that the expectation of the second coming of the Messiah is at the center of Jesus' thought, and that it dominates his feeling, his will and his action far more rigorously than we had previously supposed. At the same time, however, the ideas expressed by David Friedrich Strauss would not be silenced. Again and again the reproach was made that the portraits of Jesus, which place in the foreground what Strauss calls the quixotic and the fanatical in the world of Jesus' ideas, picture a personality with clearly revealed morbid traits.[3]

---

[1] David Friedrich Strauss, *Das Leben Jesu* (*The Life of Jesus*), two volumes (Tübingen, 1835, 1480 pp.).
[2] David Friedrich Strauss, *Das Leben Jesu fürs deutsche Volk bearbeitet* (*The Life of Jesus Revised for the German People*), (1864, 631 pp.). The statement in the letter to Wilhelm Lang is cited in Theobald Ziegler's *David Friedrich Strauss*, Part II (1908, pp. 608 f.).
[3] Students like H. J. Holtzmann and Adolf Jülicher, among others, express this point of view. See H. J. Holtzmann, *Das messianische*

In the most recent phase of the study the discussion turns almost entirely upon the question of the degree to which such ideas of Jesus may be considered authentic. Indeed, a series of attempts have been made which essentially represent the Messianic claims of Jesus and the expectation of his second coming as unhistoric. According to this hypothesis the Nazarene was a simple Jewish teacher, whose followers after his death elevated him to the rank of Messiah and then proceeded to place in his mouth allusions and expressions relating to it.[4] Nevertheless, this kind of distinction between authentic and unauthentic words in the sources cannot be maintained. It must, therefore, be admitted that Jesus considered himself to be the Messiah and expected his majestic return on the clouds of heaven. The psychopathological literature about Jesus which is here in question includes the works of Dr. de Loosten,[5] Dr. William Hirsch,[6] and Dr. Binet-Sanglé;[7] alongside of these should also be mentioned

---

Bewusstsein Jesu (The Messianic Consciousness of Jesus), (1907, pp. 80f). Theobald Ziegler gives an even more vigorous expression to these reflections, *David Friedrich Strauss*, Volume II, p. 609. See also Hermann Werner, "Der historische Jesu der liberalen Theologie, ein Geisteskranker" ("The Historical Jesus of Liberal Theology, a Psychotic"), *Neue kirchl. Zeitschrift*, XXII, 1911, pp. 347–390.

Among the non-medical books on Jesus' mentality should also be mentioned the works of Oskar Holtzmann, *War Jesu Ekstatiker?* (*Was Jesus an Ecstatic?*), (1903, 143 pp.), and Julius Baumann, *Die Gemütsart Jesu*, (*The Character of Jesus*), (1908, 90 pp.).

[4] The most ingenious attempt of this sort was by William Wrede in his well-known study, *Das Messiasgeheimnis in den Evangelien*, (*The Messianic Secret in the Gospels*), (Göttingen, 1901, 266 pp.).

[5] George de Loosten (Dr. Georg Lomer), *Jesus Christus vom Standpunkte des Psychiaters* (*Jesus Christ from the Standpoint of the Psychiatrist*), (Bamberg, 1905, 104 pp.).

[6] William Hirsch, *Conclusions of a Psychiatrist* (New York, 1912). Translated into German under the title, *Religion und Zivilisation vom Standpunkte des Psychiaters* (*Religion and Civilization from the Standpoint of the Psychiatrist*), (Munich, 652 pp.). "Ueber Jesus" ("Concerning Jesus"), pp. 87–164.

[7] Charles Binet-Sanglé, *La Folie de Jésus* (*The Dementia of Jesus*). Vol. I (3d Edition, Paris, 1911, 372 pp.): *Son hérédité; sa constitution; sa physiologie* (*His Heredity; His Constitution; His Physiology*). Vol. II (3d Edition, Paris, 1910, 516 pp.): *Ses connaissances; ses idées; son délire; ses hallucinations* (*His Knowledge; His Ideas; His Delirium; His Hallucinations*). Vol. III (1st Edition, Paris, 1912, 536 pp.): *Ses facultés intellectuelles; ses sentiments; son procès* (*His Intellectual Faculties; His

the study of Emil Rasmussen, Ph.D., who has doubtless had medical counsel.[8]

Here follows a brief summary of the principal contents of these writings.

De Loosten arrives at the following conclusion.[9] Jesus, he insists, is evidently a hybrid, tainted from birth by heredity, who even in his early youth as a born degenerate attracted attention by an extremely exaggerated self-consciousness combined with high intelligence and a very slightly developed sense of family and sex. His self-consciousness slowly unfolded until it rose to a fixed delusional system, the peculiarities of which were determined by the intensive religious tendencies of the time and by his one-sided preoccupation with the writings of the Old Testament. Jesus was moved to express his ideas by the appearance of John the Baptist. Proceeding step by step[10] Jesus finally arrived at the point of relating to himself all the Scriptural promises, which had become vital again through national misfortune, and for whose ultimate glorious fulfillment all hearts hoped.

Jesus regarded himself as a completely supernatural being. For only so and not otherwise can man understand his behavior when he arrogated to himself divine rights like the forgiveness of sins.[11]

That he kept the Messianic dignity which he claimed as much as possible to himself, de Loosten explains psychologically[12] by the reflection that Jesus did not believe he had a large enough following at that time to enable him to realize his claims.

Feelings; His Trial). Vol. IV promises: *La morale et les actes; groupement et comparaison des symptômes* (*Morale and Action; Classification and Comparison of Symptoms*). [Published later, Paris, 1915.]

[8] Emil Rasmussen, *Jesus. A Comparative Study in Psychopathology.* Translated into German by Arthur Rothenburg under the title, *Jesus. Eine vergleichende psychopathologische Studie*, Leipzig, 1905, 166 pp.
[9] De Loosten, pp. 90 ff.
[10] *Ibid.*, p. 34.
[11] De Loosten quotes Matt. 9:2; Mark 2:5–12; Luke 5:20 and 7:48.
[12] De Loosten, p. 48.

From his words to the young man who wished to attend to the burial of his father, "Follow thou me and let the dead bury their dead" (Matt. 8:22), as well as from other words, de Loosten infers, on the one side, that Jesus takes it for granted that the beginning of his divine Utopia was immediately imminent, and on the other side, that he was no longer conscious of his human nature. The journey to Jersusalem is described as a foolhardy idea [13] of achieving by a certain stroke of violence his long cherished, and a thousand times expressed, claims.

After the moment of depression in Gethsemane, his psychosis erupts at the advent of the police in all its old strength.[14] This mental disorder finds expression during the examination before the high council in which Jesus holds out the prospect to the high priest that his judges will see him sitting at the right hand of God as the Son of Man and coming on the clouds of heaven.[15] Finally, from John 7:16–20 still another idea of persecution is adduced.[16]

Concerning his emotional life de Loosten says that Jesus' temperament was not at all uniformly serene on every occasion, and that sometimes he was liable to strange and apparently groundless moods of depression.[17] In illustration the Fourth Gospel is especially cited.[18]

Before his arrest Jesus found himself in a highly nervous, excitable state. He knew what a risky game he played and suffered greatly under the weight of fears and ominous misgivings. The completely senseless cursing of the fig tree, also becomes intelligible only as springing from this mood.[19]

---

[13] *Ibid.*, p. 72.
[14] *Ibid.*, p. 83.
[15] *Ibid.*, p. 85.
[16] John 7:16–20: in a speech in Jerusalem Jesus reproaches his hearers very abruptly for seeking his life; they are astonished, and accuse him of being possessed.
[17] de Loosten, p. 65.
[18] John 12:27 is given as the principal reference.
[19] de Loosten, p. 77 ff.

The way in which Jesus here takes out his ill-humor on a defenseless tree is, as we have said, to be explained only as the boiling over of severe spiritual excitement.

The driving of the money-changers out of the temple de Loosten describes as a shocking act of violence.

Among hallucinations he mentions the occurrences at the baptism by John,[20] a vision which obviously exercised a decisive influence upon Jesus' later decisions.

This is a matter of hallucinations in the visual and auditory realms, which here certainly, as is often the case, accompany a greatly excited mind.

With what frequency Jesus had these hallucinatory visitations, we do not know, says de Loosten. He considers it probable that Jesus depends upon them even for his decisions and that similar visions like those at the baptism occurred later.

Besides the visual hallucinations, de Loosten thinks that it is highly probable that Jesus suffered from voices which seemed to him to come out of his own body.[21] Jesus placed an exalted value upon the supernatural spirit ($δαιμόνιον$) allegedly residing within him. A daimonion determined what he should do and leave undone, and he obeyed.[22]

The utterance of Jesus, which he takes out of the traditional text, "Someone has touched me, for I feel that power has gone out from me," [23] de Loosten explains in this way, that Jesus had felt some kind of abnormal peripheral sensation, perhaps of the skin, and that he was trying to find an explanation for it.

The lack of sex-consciousness which is thought to be proved in the words about the eunuch (Matt. 19:12), is

---

[20] *Ibid.*, p. 36.
[21] *Ibid.*, p. 64.
[22] *Ibid.*, p. 73.
[23] Mark 5:27–34. In reality Jesus only asserts that someone has touched his clothes. It is a naïve conjecture of the Evangelist that he said this because of a feeling that power had gone out from him.

brought forth along with the already mentioned lack of family loyalty as a sign of psychic degeneration *par excellence*, which readily fits into the previously formed picture of his personality.

William Hirsch makes a diagnosis of Jesus, namely, paranoia. Everything that we know about him conforms so perfectly to the clinical picture of paranoia that it is hardly conceivable that people can even question the accuracy of the diagnosis.[24]

Hirsch traces the development of the delusion in this way.[25] We find a boy with unusual mental talents who is, nevertheless, predisposed to psychic disturbances, and within whom delusions gradually form. He spends his whole leisure in the study of the Holy Scriptures, the reading of which certainly contributed to his mental illness. When at the age of thirty he first made a public appearance, his paranoia was completely established. It is apparently one of those cases, Hirsch believes, where sudden and formless psychotic ideas are, indeed, present, but where, nonetheless, they need an external shock and a strong emotion, in order to form a typical systematic structure of paranoia.

This shock was provided for Christ by another paranoid, no other than John the Baptist.[26] Meantime Jesus' delusions attained their most complete maturity, and when he heard of the "forerunner of the Messiah," who was baptizing sinful people in the river Jordan, he betook himself there in order to receive baptism himself. The hallucinations which appeared on this occasion are later discussed.

After the baptism Jesus went into the wilderness for forty days. This sojourn is for us of the greatest interest for these forty days lie between two sharply differentiated sections of his life. The delusions which up to that time were

---
[24] Hirsch, p. 99.
[25] *Ibid.*, p. 125; p. 100.
[26] *Ibid.*, p. 101 ff.

isolated and unrelated to each other henceforth merged into a great systematic structure of delusions; doubtless Jesus had at that time repeated conversations with God the Father who had commissioned him and whose doctrine he preached. Such a development of his illness, a transition from the latent to the active stage of paranoia, is quite characteristic of this psychosis.

In the great drama of the public ministry of Christ stretching over three years, the megalomania, which mounted ceaselessly and immeasurably, formed the center around which everything else turned.

All his sayings, his teachings, his sermons culminated in a single word: "I." Hirsch cites for this view a series of references in the Gospel of John.[27] At the conclusion of this exposition he goes so far as to assert that no textbook on mental diseases could provide a more typical description of a gradually but ceaselessly mounting megalomania than that afforded by the life of Jesus.

Ideas of reference are also found in Jesus by Hirsch,[28] in so far as he believed that all the predictions of the prophets applied to him and that he was the king who should rule over the world. Therein the Nazarene manifests one of the actual peculiarities of paranoids, who apply to themselves everything possible that they see or read.

Jesus' claim to be of the family of David, Hirsch relates to the well-known tendency of youthful paranoids to substitute for their real descent a highly colored fanciful one.

The cursing of the fig tree is represented as the action

---

[27] John 6:29, 35, 38, 40, 47–58; 7:38; 8:12; 11:25 ff; 14:6, 13, etc. It is noteworthy that he cites in this connection the words put into the mouth of the resurrected Jesus, "All power is given unto me in heaven and in earth" (Matt. 28:18). He also mentions in reference to this the words about the "Son of Man" from the first two Gospels without noticing that Jesus does not there speak of his "I," as he does not at all identify his person in these sayings with the Son of Man.

[28] Hirsch, p. 126.

of a paranoid without, however, examining this idea more closely.

Concerning the baptismal hallucination he remarks "The aberration which had so long filled the mind of Jesus, that he was the Son of God and that God had ordained him to be the Savior of mankind, was from now on converted into visual and auditory hallucinations," and concerning the following forty days of solitude he remarks: "During the forty days in the wilderness Jesus must (!) have been in a state of continual hallucination. His sojourn there must have involved hallucinations." [29] All the utterances of Jesus to the effect that he had received from God everything which he proclaimed—passages in John are partly in mind—are understood as spoken with reference to preceding auditory hallucinations.

The story of the transfiguration of Jesus on the mount (Mark 9:2–8) Hirsch also would understand as a hallucinatory experience. As a typical example of an illusion it is cited that, according to John 12:28 f, Jesus heard a voice from heaven which proclaimed his coming glorification, while the people heard a bolt of thunder.[30] Binet-Sanglé likewise advances a diagnosis, namely "religious paranoia." According to rule he distinguishes three stages:

1. The period of conception and of systematization
2. The hallucinatory period
3. The period of personality change

He discusses separately, so far as possible, the delusions and the hallucinations.

According to Binet-Sanglé, the primary delusion (the primordial fixed idea) appears *ex abrupto*, without previous reflection.[31] The further development of the delusion is apparently coherent, and though proceeding from a false hypothesis is thoroughly logical in its consequences. It de-

---

[29] *Ibid.*, p. 101.   [30] *Ibid.*, p. 110.   [31] Binet-Sanglé, I, 269.

velops by the progressive extension of the primary idea but without undergoing any transformation and without losing its original stamp.[32]

Through the suggestive power of various incidents, through John the Baptist, through his own miraculous cures, through the marveling of those who were healed of their diseases and through the enthusiasm of the disciples, Jesus is brought to the point of believing himself to be the Messiah, the King of the Jews, the Son of God, God's interpreter, God's witness, and finally of identifying himself with God. The threats of the fanatical Pharisees and Scribes also awakened in him the notion that he was the sacrificial lamb which by its death was to take away the sins of Israel, and that after his resurrection he would ascend into the heavens, there to be revealed in his complete glory.

The hallucination at the Jordan baptism is described by him as having the character of an initial hallucination coming from above[33] and of an encouraging nature. A spoken component is associated with the visual. The voice said, "This is my beloved Son, in whom I am well-pleased." This was a verbal and auditory hallucination.

According to Binet-Sanglé the flight into the wilderness followed upon the baptism. Here, under the influence of protracted abstinence and loneliness, of the quiet and the monotony of the wilderness which placed him at the mercy of all his obsessions, perhaps also under the added influence of weariness and heat, multifarious mental disturbances took form. Binet-Sanglé finds seven hallucinations in all in the account, two purely visual and five which are described as both visual and auditory-verbal.[34]

---

[32] *Ibid.*, II, 278.
[33] *Ibid.*, I, 349 ff. P. 350: "It has been noticed that, in religious paranoia as well as in hysterical ecstasy, the object of the visual hallucination almost always appears to have a certain exalted character."
[34] Binet-Sanglé, II, p. 392, "On the other hand, none of the hallucinations of Jesus is wholly verbal. For, in religious paranoia, it is very rare

The content of the hallucinations always refers to religious objects, particularly to the devil. They may be separated into the fearsome and the comforting which appear in opposition. The comforting ones are visual: at the baptism the dove appears, in the wilderness the angel of God appears, in Gethsemane Jesus is strengthened by an angel according to the text in Luke.[35]

As causes of the hallucinations (!) Binet-Sanglé stresses the excitement, assisted by the night, the solitude and abstinence.[36]

The recorded hallucinations of Jesus, according to Binet-Sanglé, cannot have been the only ones, as insane mystics almost always suffer from hallucinations of muscle-sense. "In later periods," he goes on, "come the secondary psychomotor symptoms, constituting a kind of theomanic possession." He cites as examples of sensory hallucinations the places in the four Gospels, in which Jesus says that the Father speaks through him, but concedes that they cannot be defined specifically.

Let it be further mentioned that Binet-Sanglé wishes to establish the secretiveness of the paranoid. He adduces as evidence of this the fact that the Nazarene regarded his Messiahship and certain points in his teaching as secrets to be veiled, gave evasive answers to questions and was brought to admit his system of delusions only under the stress of emotion, as, for example, in the proceedings at the trial.[37]

De Loosten, Hirsch and Binet-Sanglé busy themselves with the psychopathology of Jesus without becoming fa-

---

(?) for verbal hallucinations to appear alone without the conjunction of visual hallucinations."

[35] Luke 22:43

[36] Binet-Sanglé, II, p. 393, "In short, the nature of the hallucinations of Jesus, as they are described in the orthodox Gospels, permits us to conclude that the founder of the Christian religion was afflicted with religious paranoia."

[37] In the above outline of Binet-Sanglé's thought a number of abstruse assertions of the author are omitted from consideration.

miliar with the study of the historical life of Jesus. They are completely uncritical not only in the choice but also in the use of sources; therefore, before we can enter into a psychiatrical discussion of their studies, we must recall what they neglected. Let us, then, set forth in brief the results achieved by the criticism of sources and by the scientific study of the life of Jesus.

With reference to the sources it is first of all to be observed that the Talmud and the extra-Biblical gospels—the latter are chiefly interested in illuminating the infancy stories—should not be considered.[38] We must omit the Fourth Gospel also, for the Jesus painted there, as critical investigation since Strauss has more and more recognized, is in the main a freely imagined personality who is designed to improve and supplement the Jesus appearing in the first three Gospels; in contra-distinction to this latter personality, a mode of thought and preaching approaching the feeling and understanding of the Greek is attributed to the personality of the Fourth Gospel.[39] The Jesus of the Fourth Gospel, in accordance with Greek dogma, knows himself to be the eternal spirit of God become flesh (Logos), and expresses this in hints and allusions which must appear to his Jewish listeners, in accordance with the design of the Fourth Evangelist, as enigmas. Moreover, he refers in long discourses to the sacraments of the baptism and the last supper, sacraments performed by the spirit and actual only after his death. In the conduct and the precepts of a literary

---

[38] De Loosten is indebted to the statements handed down in Talmudic and pagan tradition—through Celsus in the second century—for the belief that Jesus was the natural son of Mary and the Roman legionary Panthere. Upon this is based his conclusion that we have to do with a "hybrid" and a "born degenerate." Pp. 19-21, 90.

Binet-Sanglé is still more uncritical than de Loosten in the choice of the sources. He considers documents trustworthy which were long ago proved spurious—as, for instance, the so-called "Judgment of Pilate."

[39] That even to the present day there are to be found defenders of the historicity of the Fourth Gospel proves nothing against the facts that are clearly evident to every critical investigator.

personality, created in accordance with such principles, much that is peculiar, unnatural and studied, comes to light, which the psychopathologists, of course, do not fail to observe, and which in their judgments, indicate illness. Three-quarters of the matter studied by de Loosten, Binet-Sanglé and Hirsch come from the Fourth Gospel.[40]

The Gospel of Luke agrees in the main with the Gospels of Mark and Matthew. Wherever it goes beyond them it makes us a doubtful contribution, which moreover is without any great significance for the criticism of Jesus and so can be left out of consideration. It should be particularly noted that the story of the twelve-year-old Jesus in the temple, which he alone presents to us (Luke 2:41–52), cannot be considered historical for a variety of reasons.

The stories of the birth and the childhood in Matthew (Matt. 1 and 2) also belong to legend, not to history.

There remain, then, as useful sources, the Gospel of Matthew, with the exception of the first two chapters, and that of Mark. They agree with one another in their construction. Matthew goes beyond Mark in a series of valuable discourses which he alone has handed down to posterity.[41] Both Gospels date from between 70 and 90 years after Christ. They go back, however, to still older sources, a fact which may be accepted as certain. In general their reports are trustworthy, and even though here and there later misunderstandings and cloudy traditions are to be observed, the exact similarity of certain details is startling.

Concerning the early life of Jesus we know very little. He came from a carpenter's family in Nazareth and himself plied that trade. He does not seem to have taken any studies leading to an appearance as a teacher. When later he ap-

---

[40] The discourses in Matthew 24 and Mark 13 do not belong to the original source material of the Gospels.
[41] See note 40 above.

peared in his home town as a prophet, the inhabitants wondered at the wisdom of the man whom they used to know only as a carpenter (Mark 6:1–5). However, it is possible that before his public appearance he had already spent a longer time elsewhere. His knowledge of the Scriptures could have been gained from the Sabbath readings.

Four brothers and several sisters are mentioned (Mark 6:3; 3:31). We do not learn where his age placed him among the children.

That the family of Jesus on his father's side was descended from David may be considered assured, and there is nothing extraordinary about it. Among those who returned from the Babylonian captivity under Cyrus were members of the royal family. The first caravan was led by Zerubbabel of the family of David, who also played an important political role at the beginning (Ezra 3:8; Zachariah 4:6 f; Haggai 1:12–15). When the blind man in the streets of Jericho addressed Jesus as the "Son of David" (Mark 10:47 f.), and the people at the entrance into Jerusalem exulted in the coming of the "Son of David" (Matt. 21:9), this accords with the witness of Paul who in his Epistle to the Romans (Rom. 1:3) about three decades after the death of Jesus assumes as well-known Jesus' descent from the royal family.

We know nothing about the physical appearance of Jesus or about the state of his health.

The first two Gospels present no picture of the life of Jesus but report only his public appearance. This at the most lasted for about a year and it ended with a journey to the Feast of the Passover. Had such a Passover fallen in the period of Jesus' public ministry, he would have had to make a pilgrimage to the capital in order to fulfill the religious obligations required by law (Deut. 16:16). Had he not done so, his opponents would have been justified in reproaching him with laxity in the observance of the commandments, an opportunity which assuredly they would not have missed as it was a matter of concern to them to undermine his authority by all possible means. As the two oldest Gospels say nothing either about a Passover falling within the public ministry of Jesus, or of explanations of Jesus' neglect of one, it is to be accepted as established that no spring festival took place within that period.

We do not know how long a time Jesus spent among the followers of John. When John was taken prisoner Jesus came into Galilee and preached the same tidings that the prophet had proclaimed by the Jordan: "The Kingdom of God is at hand." The summons to repentance was added to this overture (Mark 1:14 f).

This proclamation remained the same from the first to the last day of his ministry. The Kingdom of God, also called "the Kingdom of Heaven," is synonymous with the "Messianic Kingdom." The proclamation of its nearness signifies that the end of the world is also near. The Kingdom of God is thought of as a supernatural period of time which is to set in after the natural order has ended. The latter is characterized by the fact that the evil angels have seized control of it, so the evil and imperfection in the present state of things are explained. Sickness and death also result from this fact. At the moment when God puts an end to this interim domination the perfect takes the place of the imperfect and the good takes the place of the evil. The earth also is then transformed into a splendid state and a new fertility takes the place of the old. At the same time death loses its power. The generations that now lie in the grave rise and with those who survive are assembled before the throne of God, where the evil and the rejected fall into eternal torment, while the chosen enter into a state like that of the angels, begin a blessed endless life and are gathered for the Messianic feast. The judgment according to most statements is delivered by the Messiah.

Jesus did not need to describe this future in closer detail. His hearers knew what it was all about, as soon as the sentence was spoken, "The Kingdom of God is at hand." The details are well established from the books of the prophets and the apocalypses (revelations). As the most significant of the latter is to be noted the book which, by a literary fiction, has been ascribed to Enoch and which in its major portion comes from the beginning of the last century before Christ. This book had a very strong influence upon the idea world in which Jesus lived. Beside it we must also consider the book of Daniel which appeared in the year 165 B.C. and has already a completely apocalyptical character.

From the literature which was then at hand it was clear to Jesus' listeners that the last days in the course of the natural world would be filled with terrible wars and unprecedented miseries for mankind. The rabbis called this time "The Woes of the Messiah."

This is the content of the gospel which Jesus preached. He proclaims the nearness of the Kingdom and of the judgment. At the same time he amplifies his message by explaining in what the repentance and morality which are necessary for justification at the judgment consist, and reminds them of the great misery which still stands before them. He who is faithless in the time of persecution is lost; he who wishes to save his earthly life will lose his eternal life. Concerning himself Jesus expects that he must at that time endure great disgrace and persecution and implores his followers not to be led astray then, and in spite of all, to remain loyal to him (Mark 8:34–9:1).

In this connection, it should be noticed that in the expectations of the late Jewish world an older prophetic way of looking at things and a younger apocalyptic way of looking at things exist side by side. In accordance with the earlier view a distinction is drawn between the Messianic

Kingdom and the splendid, definitive end-state of the world which does not come until after the former; according to the later idea, which was shared by Jesus, both occur together.

The differences between these two points of view in connection with their ideas of the future ruler, have a certain significance. The prophetic notion is that he is a shoot from the stem of David, who at the proper time will be endowed by God with supernatural power and wisdom and anointed the King of Glory. It is for this reason that he is called the Messiah (the anointed one). In the book of Daniel the consequence is seen that there is no longer any reigning family of David out of which a ruler could be raised to the position of Messiah. Therefore the author expects that God will confer this highest power in the coming world state to an angelic being who has a human form and looks like "the son of a man" (Dan. 7:13 f).

In the later apocalyptic literature, and already in the Apocalypse of Enoch, the "Son of Man" is merged with the Messiah, although originally they had nothing to do with each other. From this identification arises a curious problem. The Son of Man is an angelic being who has no human line of descent. How can he then be at the same time the "Messiah," who must be born naturally from David's family? Contemporary Scriptural knowledge, so far as we know, did not express itself on this question precisely because it had discovered the inner cleavage in these expectations of the future. Jesus, on the contrary, did concern himself with the question. His solution of the problem is of meaning to us in so far as it vouchsafes us a glimpse into the considerations from which grew the faith that he was to appear as the Messiah.

It should be said, to correct obvious misunderstandings, that Jesus does not mean that he as a normal man is already in his lifetime the Messiah, or the Son of Man. His conviction is that he is ordained to this dignity and will be revealed in it at the end of the world. This is the meaning of his words to his judges. After he has answered in the affirmative the question of the High Priest, if he holds himself to be the Messiah, he adds: "You will see the Son of Man sitting on the right hand of power (that is, God) and coming upon the clouds of heaven" (Mark 14:62). In the natural era of the world the Messiah can be no more present than the Kingdom of God itself.

Jesus did not permit the conviction that he was destined to be the coming Messiah to play a part in his message. This is one of the most certain results of modern critical research. He holds out the prospect of the immediate coming of the Son of Man but always speaks of him in the third person so that none of his listeners could get the thought that he himself expected to be clad with this honor. Until the last day he remained for the people the prophet from Nazareth. It is the latter, not the Messiah, who receives the ovation at the entry into Jerusalem. Even the Pharisees and the Scribes, who come to terms with him during the last days, have no suspicion of his claims; otherwise they would have brought it up for discussion. One is aware that the High Priest is not able to summon a single witness at the trial to testify to the Messianic claims of Jesus.

Only the disciples know something of his secret. They learn it for the first time, however, a few weeks before his death. He reveals it to them in the moment when he sets about going to Jerusalem (Mark 8: 27–30), and one of them, Judas, betrayed it to the High Council. On the basis of this disclosure Jesus was arrested the eve of the day when the Passover lamb was eaten, or perhaps on the very evening of this last day, was condemned on the same night in disregard of the laws governing capital trials, and immediately crucified. As the High Priest had only one witness for his accusation, and as, according to Jewish law, at least two were necessary before sentence could be pronounced, everything depended on whether Jesus admitted his guilt or not. He could have saved himself by silence. But he did not wish to do so as he had come to Jerusalem with the settled purpose of dying and had even disclosed this to his disciples (Mark 8:31; 9:31).

He is convinced that his death is an atonement (Mark 10:45), because of which mankind will be exempted by God from the general misery which is to precede the Messianic Kingdom, and expects that he, at the moment of his death or on the third day after his death, will enter into the supernatural life, achieve the Messianic honor, and usher in the end of the world, the judgment and the Messianic Kingdom. The disciples understand his allusions in this sense and on the way to Jerusalem quarrel over which of them will be called upon to occupy the place of highest honor beside him (Mark 9:33 f; 10:35-45).

It should be noticed also that Jesus did not work publicly during the whole period between his first appearance and his death. After he had preached for a time the nearness of the Kingdom of God he came to believe, probably at the time of the harvest, that the expected moment had come and sent his disciples out two by two to carry the tidings into the districts of Israel. As he sends them out he announces to them in a significant speech (Matt. 10) that the time of great distress will forthwith begin. Simultaneously he holds out before them the prospect that the coming of the Son of Man will take place before they have had time to carry out his commission (Matt. 10:23). He expects also in this time of stress while they are away from him to be transformed into the Messiah.

When they returned to him without his predictions having been fulfilled he withdrew with them into the heathen country. In the midst of his success he abandoned the multitudes who had gathered around him to await the coming of the Kingdom of God, and spent the time till his departure for Jerusalem (that is, the autumn and winter) in heathen territory, in the neighborhood of Tyre and Sidon and Caesarea Philippi, without preaching and intent only on remaining unknown. In this period he accounted for the non-arrival of the period of trouble and of the Kingdom of God in this way: that he as the coming Messiah was called upon to suffer and die for others. In the formulation of this thought, the death of John the Baptist (which occurred after the sending out of the disciples), in whom Jesus had recognized the Elijah who had been promised for the last days (Mal. 4:5 f; Matt. 11:13–15; Mark 9:11–13), and the 53d chapter of the book of Isaiah, which speaks of the Servant of God who suffers for the guilt of the people, were probably decisive.

Is it possible from the sketch here given, and from the many details which could not be taken into consideration to draw conclusions like those which de Loosten, Hirsch and Binet-Sanglé have reached? Already in our introduction the reasons are given which must make us highly skeptical here. Nevertheless, let us go into the substance of these writings in order to form a clear opinion.

From this hurried glance at the sources, the contents of the preaching of Jesus, and the course of his public appearance, we may gather how much of the material used by de Loosten, Hirsch and Binet-Sanglé is untenable if one deducts from it that which is not trustworthy from the point of view of historical criticism.

Nothing that is adduced from them concerning Jesus' childhood and youth, concerning his predispositions and his development can claim any validity. The story in Luke of the behavior of the twelve-year old Jesus in the temple, which so greatly interested the psychopathologists, and in which Binet-Sanglé was inclined to find the account of a hebephrenic crisis itself, must be dismissed from consideration. Discarding the Fourth Gospel is of the greatest importance. This source long permitted the psychopathologists the assumption that we can follow Jesus' mental development through the course of three years; only this allowed them to draw a personality continually occupied with his ego, placing it in the foreground of his discourses, asserting his divine origin and demanding of his hearers a corresponding faith.[42]

Because of the fact that the three psychopathologists confuse the portrait in John with that in the older Gospels—in which Jesus does not speak of himself or of his dignity—they come to the conclusion that sometimes he proclaimed himself as the Messiah, and sometimes refrained from doing so, interpreting this conduct in terms analogous to those which are properly applied to paranoids.

We cannot enter one by one into a consideration of the numerous and, in part, very grievous misunderstandings which crop up from de Loosten's, Hirsch's and Binet-Sanglé's unfamiliarity with the apocalyptic world-view of late Judaism, and with critical investigation.

In connection with the psychiatric evaluation of the material at hand we must once again point to the very extraordinary nature of the difficulties which confront every effort at a diagnosis.[43]

Since the manner and the method by which a mental

---

[42] Moreover, whatever de Loosten and Binet-Sanglé bring forward about ideas of persecution and about inexplicable moods of depression is taken only from the Fourth Gospel. So, also, that which Hirsch adduces as illusion on the part of Jesus, basing his assertions on John 12:28 ff, is to be dismissed on the basis of our judgment as to the nature of the sources. The affected and unnatural manner of speaking to which Binet-Sanglé alludes is also peculiar to the Fourth Gospel.

[43] It is to be noticed that de Loosten and Binet-Sanglé lay great stress upon the statements concerning the mental condition of Jesus which have been handed down to us from contemporaries. They appeal to the fact that his opponents among the Pharisees maintain that he is "possessed" (Mark 3:22), and that his family wish to bring him home from Capernaum to Nazareth because he is "beside himself" (Mark 3:21). From this, however, one may only infer that the former wish to discredit him with the people at any cost, and that the relations perceive

aberration is expressed are much plainer diagnostically than its content, it is immediately evident that biographical material is to be utilized with extreme caution. The majority of the "symptoms" which are contemporaneously evident are not set down in writing, but rather their verbal manifestations only. If, then, diagnostic conclusions are deduced from this kind of material, they must necessarily be for the most part of a very hypothetical kind.

When, for instance, Binet-Sanglé, Hirsch and de Loosten maintain that Jesus had numerous hallucinations during the forty days in the wilderness, even if we assume the historicity of the sojourn in the wilderness, this must be a wholly vague hypothesis although it allows them to affirm a hallucinatory phase in the development of his psychosis.

And even in the case of actual historical utterances the interpretation can very easily be an arbitrary one. So, for example, the psychological explanation which Binet-Sanglé advances to explain the origin of the hallucinations which he attributes to Jesus seems wholly artificial. To offer with assurance such a mechanical explanation of the appearance of hallucinations even in a living patient with mental disease, a many-sided and comprehensive analysis of the individual would be necessary.

The circumstances in the present case are still more strangely complicated by the fact that the three authors in question believe that we have to do with the disease of paranoia.

It is well known that the question of paranoia belongs among the most difficult problems of modern psychiatry. It is still very far from solution. It may, however, be emphasized here that the controversy now going on about

---

a change in him and are not able to explain to their satisfaction how it comes about that he sets himself up as a teacher and a prophet. Besides, it should be laid down as certain that the Pharisees and the followers of Jesus do not declare that he is beside himself because he considers himself to be the Messiah, for they know nothing whatever about this claim.

the nature of paranoia is in no small degree also a controversy over words. A considerable number of the kinds of paranoia are sufficiently well known in the way they progress to permit a profitable discussion of differential diagnosis among those who are concerned not with the words but with the facts.

Binet-Sanglé proposes the diagnosis of religious paranoia. From the general discussions included as a kind of fragment for a psychiatric textbook in his psychopathological work we may conclude that he distinguishes three phases, as Magnan does in his "chronic delirium in its systematic development" (délire chronique à évolution systématique).

Hirsch seems to have a similar form in mind. He fails, however, to characterize it sufficiently, and without further ado presumes "known paranoia."

De Loosten lays less stress on the hallucinations and concedes that the factors of physiology and genius play a more significant part in the nature of Jesus than the other authors admit.

It cannot be the purpose of this study to express an opinion for or against the existence of any particular form of mental disease in Jesus, or to discuss a clinical diagnosis. Its purpose is merely to test the elementary symptoms which the three authors have used to support their diagnosis for their historic authenticity, and in case this is established, for their clinical value.[44] The single alleged symptoms—1. Delusions, 2. Hallucinations, 3. Emotional attitude, 4. Other characteristics—will be separately treated.

---

[44] Dr. H. Schafer comes to grips with de Loosten in *Jesus in psychiatrischer Beleuchtung* (*Jesus in the Light of Psychiatry*), (Berlin, 1910, 178 pp.). He is no more a master of the historical material than his opponent. For the sake of completeness the refutations of those who are not doctors should be brought forward. Philipp Kneib, *Moderne Leben-Jesu-Forschung unter dem Einflusse der Psychiatrie* (*Modern Research into the Life of Jesus under the Influence of Psychiatry*), (Mainz, 1908, 76 pp.). Hermann Werner, *Die psychische Gesundheit Jesu* (*The Mental Health of Jesus*), (Gross-Lichterfelde, 1908, 64 pp.).

Before we begin the discussion of the picture of psychosis drawn by the three authors, attention should be called to the fact that the assertion that this is a question of acquired mental disease has already *a priori* very little probability about it. Binet-Sanglé cites a large number of clinical observations of sick people who have suffered from religious paranoia accompanied by all kinds of hallucinations and also points out the frequency with which these cases, so familiar to every psychiatrist, occur. He forgets, however, that cases of "chronic delirium in its systematic development," which resemble the paranoiac form of Kraepelin's *Dementia praecox*, are for the most part hospitalized soon after the onset of their illness, and that these forms of mental disease are exactly the type which do not win supporters and disciples and found sects. The numerous hallucinations, the catatonic symptoms in the broadest sense of the word (autism), and the effects of dissociation make these sick people incapable of consecutive activity; if some renunciation of activity takes place it seldom corresponds to the delusional contents of a consciousness which has been impaired by psychosis. The fact is well known that sick people suffering from delusions of persecution, particularly from a morbid fear of physical persecution, continue to carry on their work years on end and extremely seldom draw the practical conclusion from their hallucinations and delusions, namely that they should defend themselves against their persecutors—either legally or illegally; if in some fleeting mood they do this, it happens because of some state of excitement, and not, therefore, because of conscious and logical inferences. The persecuted psychotics, who are actually and continually on the defensive, the "persecuted persecutors" of the French, belong to the group of congenital psychopaths and do not suffer from an acquired mental disease.

In the examination of that which is alleged to be delusion

in Jesus, the origin of the delusion is to be distinguished from its later development.

De Loosten assumes a slow development. "Jesus Christ," he writes, "worked up to a fixed system of insanity by slow development." The appearance of John the Baptist provided the first impulse towards the externalization of his ideas.

Hirsch is of the opinion that we have to do with a case in which unsystematized and undefined delusions were present, in which, however, these ideas needed an external shock and a strong emotion in order to take shape in a typical psychotic system.

Binet-Sanglé maintains that the primary delusion (the primordial fixed idea) appeared *ex abrupto*, without any previous consideration.

No one of the three authors states clearly what the content of this primary delusion was. From their other statements we may conclude that Jesus' belief in his own Messiahship should be considered as the primary delusion.

As for the origins alluded to, they are certainly not to be thought of as proved since we are not at all acquainted with the period of Jesus' life that precedes his public appearance. As de Loosten and Hirsch do not further specify the character of the paranoia, we cannot discuss at all the probability of the origin accepted by them.

Binet-Sanglé in his assumption of the appearance *ex abrupto* of the primary delusion affirms an origin which is not usually found in the form of the disease diagnosed by him. In the first period, which Binet-Sanglé often confuses with the second hallucinatory period, a strong idea of reference (the insanity of interpretation) usually appears; at first the delusion develops gradually, showing constantly a persecutory character which absolutely cannot be brought into harmony with any idea of the Messiahship.

In another place Binet-Sanglé on the contrary emphasizes that Jesus was brought to think of himself as the

Messiah by the suggestive influence of several events: by John the Baptist, by his own wonder cures, by the astonishment of those who were healed and by the enthusiasm of the disciples. He attempts in this place, then, a psychological explanation.[45]

So far as the further development of the delusion, the so-called "systematization," is concerned, all the authors account for it in the sense of an expansive tendency.[46] Binet-Sanglé entitles the chapters dealing with this matter progressively: the Messiah-king; the Son of God; the Confidant, Interpreter, Agent of God; Jesus-God.

De Loosten emphasizes the fact that Jesus, proceeding step by step, arrives in the end at the point of referring to his own person all the promises of the Scriptures and their final glorious fulfillment.

Hirsch writes that after the forty days in the wilderness the theretofore isolated and disconnected delusions unite in a great and systematic structure, and asserts that then, in numerous hallucinations which arose at the same time, Jesus emerges from the latent to the active stage of paranoia. In another place he speaks of a gradually evolving and ceaselessly mounting megalomania.

So far as the development of the psychosis which is accepted by the three authors is concerned, we should emphasize that de Loosten speaks of a continually mounting curve, while Binet-Sanglé and Hirsch regard the forty days

---

[45] In just the same way, he tries to explain the hallucinations by a large number of physiological and psychological reasons, although these, in accordance with his own diagnosis, are rather to be comprehended as emerging primary irritations in the evolution of "chronic delirium," and so have their roots in the preceding stages of the illness.

[46] Hirsch and de Loosten speak of the ideas of reference which begin to find expression here, in so far as Jesus has arrogated to himself the Messianic functions of the prophets. In any just historical evaluation of his point of view, this interpretation of passages in the Old Testament, so far as we can control them in the case of Jesus, presents a thoroughly normal psychological activity, which does not permit us to speak of ideas of reference.

in the wilderness as a dormant period and propose a later development as connected with it. No one of the authors speaks of a period filled with ideas of injury and persecution, which experience shows is characteristic of the first phase. Only Binet-Sanglé mentions the hallucinations about the devil in the wilderness, which are not pictured as entirely expansive.

It would be necessary, provided the assertions of the three authors were to the point, to consider the infrequent appearance of such a one-sided development of megalomania. It is well known that the mentally sick can evince delusions of grandeur through a long period of time,[47] but in this case no progression is present and so there is no question at all of that "systematic evolution" which the three authors seem to have in mind.

Now that it has been shown that the evolution of mental disease assumed by the three authors is quite improbable from the general clinical standpoint, it remains to discuss the question of whether the words and actions of Jesus, which even historical criticism accepts as authenticated, can provide a basis for thinking of any pathological distortion of the content of consciousness.

We must consider here, first of all, that the ideas of religion which Jesus shares with his contemporaries and which he has accepted from tradition may not be considered as diseased, even when they appear to our modern view entirely strange and incomprehensible. De Loosten, Hirsch and Binet-Sanglé repeatedly transgress this fundamental rule.

To these accepted ideas belong: the view that the present world has been given over to the evil spirits, who can even take possession of men and speak through them; the expec-

[47] Compare the expansive form of paraphrenia: E. Kraepelin, "Zur paranoiafrage" ("Concerning the Question of Paranoia"), *Die Zeitschrift f. d. gesamte Neurologie und Psychiatrie (The Journal of Neurology and Psychiatry)*, 11:617-638, Sept. 1912.

tation that the good angels led by the Messiah will take over the rule of the world; the belief in the coming of the Messianic Kingdom with all the subsidiary ideas—the misery that is to precede the Messiah, the judgment, the annihilation of the wicked, the resurrection and "transformation" of the just and the elect, the transfiguration of earthly nature into a state of supernatural fertility—which appear in the prophetic books and in the apocalypses.

The Messianic expectations belong to the stock of late Jewish dogmas. But not all the Jews believed that these events were immediate. This conviction was widespread within a movement specifically Jewish which had its origin in John the Baptist.

Concerning the details of the Messianic ideas, a certain variety of views closely approaching each other was still prevalent at that time.[48] It is not surely attested in the apocalyptical and rabbinical sources for the time of Jesus that the Messiah must suffer. On the other hand, this idea became more possible as soon as men let the Messiah experience an anonymous earthly existence before his appearance in glory, since he then had to suffer with the chosen the misery that preceded the Messiah. The Messianic meaning of Isaiah 53, which speaks of the suffering servant of God, could also implement the conception of the suffering Messiah.

So far as the material we have received permits us to draw a conclusion, we may say that it was not foreseen in the world of ideas which made up the late Jewish religion, that the very man who should appear as the Messiah in glory would first live an anonymous earthly life in the earthly world, since far-reaching uncertainties and unsolved contradictions dominated the basic questions about the personality and the appearance of the Messiah.[49]

---

[48] See pp. 49 f. above.

[49] According to the oldest prophetic view the Messiah was to be born as a human being in the family of David and then was to be endowed by God with supernatural attributes.

At the moment, however—this conclusion is of the greatest importance for an insight into the thought world of Jesus—when a late Jewish thinker undertook to comprehend and to clarify the problems of the Messianic teachings, and immediately resolved to hold fast to the Messiah's descent from David which the prophets had insisted on, no other solution was possible than that he should be born in the earthly era of the world from among the descendants of David and then should be raised to the supernatural splendor of the Messiah and Son of Man, through the transformation which was ordained for all of the elect at the end of the world. If the man in question was of the opinion that the expected events were near at hand, he would have to look for the future Messiah as already among the then living descendants of David.

It should be added that when the Jewish Messiah is called at the same time the Son of God, this has nothing whatever to do with descent from God in any metaphysical sense. The Son of God is only a title that indicates that his place of honor originates in God. In this sense the Jewish kings were already the Sons of God.

If after this hurried survey of the late Jewish worldviews, an analysis should be attempted of the attitude which Jesus himself assumed in relation to this circle of ideas, one would have to discuss from the psychiatric standpoint, first, his initial attitude to the Messianic idea especially, and then the further development of his personal comprehension of it.

The fact that Jesus regarded himself as the coming Messiah must strike one, but most of all has aroused the suspicion of some existing mental disturbance. De Loosten, Hirsch and Binet-Sanglé have not failed to bring forth the glaring antithesis between the attitude of the carpenter's son and that of the Messiah and Son of Man. This contrast, however, is not so lacking in sense as at first sight it appears to be if one leaves out of consideration the ideas implanted by history and by the late Jewish beliefs. In the first place, according to the oldest tradition Jesus is really from the family of David; therefore it is not possible to explain the pertinent accounts, as, for instance, Hirsch explains them, by assuming that Jesus, in common with youthful paranoiacs, has altered his ancestry in a grandiose way as a result of delusion.

In the second place, it was foreseen in the apocalyptical expectation as it has been brought out, that all the chosen would be changed at the coming end of the world into supernatural beings and in such a way, indeed, as to constitute a reversal of values. Those who in this world were lowly and despised, who had to serve and who were persecuted, would there be honored, while those who were honored here, who stood high, who ruled and enjoyed power, would come before the judgment. Jesus shared this apocalyptic view of late Judaism at all points as is evident in a succession of sayings. This view demanded exactly this, then, that he who was to occupy the highest place in the future world should here in the natural course of the world belong to the despised and common people. Therefore, only a descendant of David living in lowly poverty could be considered as the possible future Messiah. If the end of the world were thought of as being very near, then it must be that this man had already been born and had to be sought within the generation which was to experience the end of the era.

The attitude of Jesus is up to a certain degree motivated by the considerations just brought forward, which from the point of view of late Judaism are thoroughly obvious. None the less the very fact that he regarded himself as the man who would enter upon the supernatural inheritance of the family of David, remains still a striking thing.

A psychological analysis of this attitude is not possible for reasons already discussed. We can only say so much, that the exaggeration of an idea does not in itself justify our considering it the manifestation of a psychosis.

The only possible proof in the present case, that we have here in the Messianic claims of Jesus a morbid belief, would be the further fate of this idea, in other words the evidence of a mental disease with which it was associated.

The question arises here whether the change in the views of Jesus which can be proved to have occurred during the year which is pertinent in this connection shows symptoms which remind us of the evolution of mental disease. The facts which were brought forth in our survey of the public ministry of Jesus and his world of ideas need not be repeated here. Only this much needs to be emphasized, that ideas of injury and persecution never arose, and particularly did not occur in the earliest period of which we know anything about Jesus, although our present view of a "systematic evolution" would demand this.

Likewise nothing can be firmly established about any transformation of the mental disease arising from inner causes. The changes in Jesus' ideas are always conditioned by outward circumstances and represent completely logical consequences in harmony with the total picture. The delay in the arrival of the expected Messianic misery and of the Messianic Kingdom after the disciples were sent out, as well as the death of John the Baptist which occurred at that time, produce a change in the notion of suffering, because of which Jesus no longer assumes as theretofore that he is to suffer together with the chosen ones (his supernatural character already evident) the "woes of the Messiah," but rather assumes that by virtue of his suffering the others will be spared the suffering they were to have gone through.

This modification of his views presupposes a susceptibility to influence which does not accord with the forms of paranoia which develop in accordance with a firmly established type.

Moreover, the factor of antagonism is lacking, as we have already emphasized in the discussion of the one-sided, expansive evolution of the mental disease assumed by the three authors. Jesus had, indeed, enemies and opponents because he spoke out against the narrow-minded and external piety of the Pharisees. In relation to these opponents, not imaginary but genuine, Jesus conducts himself in a fashion diametrically opposite to the conduct of a sick man with a persecutory trend. He does not remain inactive and does not limit himself to a defensive attitude like so many of the sick who believe themselves persecuted, but rather seeks by actions which have a provocative character—the driving from the forecourt of the temple of the traders and money-changers, the discourses against the Pharisees (Matt. 23)—to bring on a conflict with the authorities and to force them to take steps against him, until in the end he brings the high council to the decision to get rid of him even before the festival.

This conscious effort to achieve his death can in no way be described, as Binet-Sanglé seems inclined to describe it, as morbid self-sacrifice, and cannot be brought into harmony with the corresponding actions of the mentally diseased—of which the French writer produces a considerable number of instances. As already mentioned, this sacrificial death represents a necessary constituent part in the Messianic thought and action of Jesus.

Concerning the question of hallucinations it should be remarked at the outset that de Loosten, Hirsch and Binet-Sanglé go far beyond the recorded facts here. They assume that Jesus had frequent and very extensive hallucinations and find special support for this in the utterances of Jesus reported in the Fourth Gospel, which they interpret to this end by doing great violence to the material. Because of the unhistoric nature of this source everything that was thought to be present there in the way of Jesus' hallucinations is untenable. Even the hallucination in Gethsemane, into the details of which Binet-Sanglé enters—an angel from heaven is said to have appeared to Jesus (Luke 22:43)—cannot be considered. As a comparison of the Gospel of Luke with both the older Gospels shows, we are dealing here with a legendary elaboration of the scene which precedes the arrest.

The assumption of the three authors that the forty-day sojourn in the wilderness was filled with numerous hallucinations rests upon no textual foundation, as has already been mentioned. But the three temptations of which Matthew (4:1–11) informs us are also unhistorical; they belong to the prehistoric legend as David Friedrich Strauss has already rightly remarked. The whole wilderness episode is to be evaluated on the whole as a literary product that grew out of reasoning based on the Old Testament. As Moses had spent forty days in solitude before the giving of the law (Ex. 24:18), so Jesus also must have done this before he took up his office. And as the wilderness is thought to be the residence of the evil spirits, he must have been tempted by them.[50]

In the story of the transfiguration on the mount we have to do not at all with Jesus' hallucinations but with the hallucinations of the three close disciples who are with him, as is clearly shown in the account (Mark 9:2–8). While they tarried with him, he was, we are told, transfigured before them. His garments appeared to them to be gleaming white and they saw Elijah and Moses coming to him and talking with him. Afterwards they heard a voice out of the clouds, "This is my beloved son, hear ye him," and suddenly, look-

---

[50] The process which has led to the appearance of these stories can still be plainly followed. Mark makes the general statement that Jesus spent forty days in the wilderness, tempted by the devil and served by angels (Mark 1:12 f). Matthew fills in this framework with detailed stories.

ing around, they saw no one with them but Jesus alone.[51]

It is also doubtful if the hallucinations at the baptism are historical. We must again and again make it clear to ourselves that the Nazarene comes into the light of history for the first time on that day when he appears as a preacher in Galilee and that everything that comes before that belongs to dark and uncertain tradition. It is possible that Old Testament reasons account for the origin of the story of the baptism. The voice from heaven sounds remarkably like the seventh verse of the second Psalm which is usually interpreted in a Messianic sense. It is extraordinary also that Jesus when he makes known to his disciples his Messiahship (Mark 8:27-30) does not mention that he was called to this place of honor at the time of his baptism, and that Paul also does not refer to the baptism of Jesus at all.

Assuming that the hallucination at the Jordan is authentic, one may say from the standpoint of psychiatry so much, that it has to do with coherent hallucinations in both the visual and the auditory sphere, that the content of the hallucinatory language is highly colored and that it reminds us in its style of words from a Psalm which is usually explained as Messianic.

Emotionally colored hallucinations—even Binet-Sanglé recognizes this—are not to be found only in the mentally diseased. They appear also in individuals who are very excitable emotionally, but who nevertheless can still be considered as falling entirely within the category of healthy people. Moreover, we should reflect to what a great extent the reading of the prophetic visions described in the Old Testament, the ideas of late Judaism in particular, and the

---

[51] Apparently in the earliest tradition of this scene it was a matter only of Peter's hallucinations which afterwards are attributed by the later reports to both of the others because they are there also. We remember that it is Peter, also, who has the first vision of Jesus after his death.

great excitement which rose with the expectation of the immediate coming of the end of the world, facilitated the rise of hallucinations in individuals predisposed to them. In any case the appearance of this one hallucinatory event, even if we wish to regard it as historical, does not permit us to conclude that a mental disease exists.

In regard to his emotional attitude de Loosten and Binet-Sanglé maintain that there is instability of a morbid kind in the feeling of Jesus. First of all, it is necessary here to inquire if emotional instability can be substantiated in him at all. It may, of course, be affirmed that periods of accentuated activity alternate in Jesus with other periods when he avoids the people and tries to withdraw into the solitudes. He displays no consistently ordered activity. His public ministry dissolves into a rather disordered running to and fro, in which he now appears on the east bank and now returns to the west bank, until in the end he takes himself off to the solitude of the north.

This alternation between the public activity and the quest for solitude is at first so puzzling that one might be tempted to confuse it with emotional instability. The account of the first entry into Capernaum is very enticing for the psychopathologists in this connection (Mark 1:21–39). It is related that after he has spoken in the synagogue in the daytime, and has healed many people in the evening, and has disappeared in the night and has been found in the morning by Peter and his comrades in a lonely neighborhood, he would not let himself be persuaded to return for a time to the city where everyone was waiting for him.

But this evasion and similar later efforts have nothing to do with emotional instability. The explanation is simply this, that Jesus for certain reasons which are still in part perceptible to us sought to avoid gatherings of people. One of the principal reasons is that he wished to shun the people because they brought the possessed and the sick from all sides in order that he might heal them. The encounter with the former especially was extremely distasteful to him as appears from significant details in Mark's account (Mark 1:34; 3:12).

Binet-Sanglé and de Loosten are not able to bring forward many other morbid symptoms. In the main they depend upon the lack of any family loyalty which they believe they have discovered in the behavior of Jesus to his relatives, in a succession of utterances, and in the lack of sexual feeling for which they bring forward the words about the eunuch (Matt. 19:10–12).

Jesus takes a hostile attitude towards his family because his relatives wish to take him home and to obstruct his public ministry (Mark 3:21). When, moreover, he declares that the bonds knit between men by the common faith in the nearness of the Kingdom of God are holier than the ties of blood (Mark 3:31–35) and desires that men in the days of the coming persecution should not be led astray by looking back toward their relations, this is not a lack of family loyalty to be accounted for psychopathologically, but a special point of view to be explained by peculiar preconceptions contemporarily conditioned, as indeed everything is to be explained by those premises which the psychopathologists consider moral defects in the ethical conduct and teaching of Jesus.

The word about the eunuchs and those who have made themselves such for the sake of the Kingdom of God is striking. The psychopathologists are, however, in error, if they want to get from this the idea of morbid sexual feeling. They do not notice that shortly before (Matt. 19: 3–9), Jesus has spoken about marriage in a very natural and affirmative way.

The key to the much discussed and much misunderstood saying is offered by passages in the Old Testament and late Jewish writings. In Deuteronomy 23:1 it is ruled that the eunuchs must be excluded from the religious community. In a later post-exilic prophetic book that was appended to the book of Isaiah and is characteristic of the generous attitude of that period, it is promised to them that they will receive a splendid reward from the Lord for their keeping of the law and the Sabbath, and in compensation for their lack of posterity will not only be made equal to the others in the expected future but will even be set above them (Is. 56:3–5). The Wisdom of Solomon, a book which appeared in the first century before Christ, speaks in a similar way (3:13 f).

The opinion of Jesus moves in the same direction. He sees in the eunuchs the despised ones who like the children are destined to honor in the Kingdom of God because formerly they had been among the rejected ones. In connection with this he makes the mysterious surmise that men have placed themselves in the class of these despised ones in order to participate in that special future honor. This case has nothing whatever to do with him as he cherishes a very special expectation of a high position in the Kingdom of God through his descent from David. The words, then, have nothing to do with sex feeling but are to be explained by the ideas found to be present in late Judaism.

There still remains to be discussed the apparently senseless act of cursing the fig tree (Mark 11:12–14). The story has come down to us in a later setting as is evident from the observation that on the following day the tree has already withered (Mark 11:20).[52] We must, however, retain without question the historical kernel that Jesus pronounced the curse over a fig tree on which he had vainly sought to find nourishment that no one thereafter for all eternity

---

[52] The story of his calming the storm on the lake (Mark 4:35–41) is not considered because it has already taken entirely the form of a great nature miracle, and the historical kernel is no longer clearly visible.

should find fruit on it. He condemns it, therefore, not to withering but to unfruitfulness for all eternity.

The deed, however strange it may seem to us, is not at all senseless in the light of his preconceptions. The late Jewish apocalyptic literature expected that even nature would participate in the transformation and would become capable of a wonderful fertility. So the Apocalypse of Baruch tries to imagine the future yield of a single grapevine (Apoc. Bar. 29). As Jesus expected the Messianic kingdom as something very near, his words about the tree refer to its fate in the new era of the world. While all other vegetation will achieve a wondrous fertility this tree is to remain barren because it deceived by the richness of its foliage the unrecognized future Messiah in his earthly humility and hunger. Since after the end of the world the Messiah is the ruler of all the creation, Jesus simply pronounces a judgment here whose execution, he believes, lies within the scope of his future power. This is not, therefore, a question of senseless rage against the natural order but a kind of advance wielding of powers which he exercises in other respects. So he decides in advance that the twelve disciples at the last judgment will have jurisdiction over the twelve tribes of Israel (Matt. 19:28); so he pronounces a curse over the Galilean cities which remain unrepentant (Matt. 11:20–24); so he condemns Jerusalem to become a waste (Matt. 23:37–38).

In these and similar words we have to do with promises and judgments which he thinks to carry out as soon as he is established in his Messianic power. The saying about the fig tree is one of a whole series of such utterances. The more they strike us as remarkable, the more understandable they are from the point of view of Jesus' late Jewish ideas. Binet-Sanglé and de Loosten discover the morbid here and everywhere where it is not present simply because they do not realize this.

The criticism of the psychopathological writings which we are considering yields, then, the following results:

1. The material which is in agreement with these books is for the most part unhistorical.

2. Out of the material which is certainly historical, a number of acts and utterances of Jesus impress the authors as pathological because the latter are too little acquainted with the contemporary thought of the time to be able to do justice to it. A series of wrong deductions springs also from the fact that they have not the least understanding of the peculiar problems inherent in the course of the public ministry.

3. From these false preconceptions and with the help of entirely hypothetical symptoms, they construct pictures of sickness which are themselves artifacts and which, moreover, cannot be made to conform exactly with the clinical forms of sickness diagnosed by the authors.

4. The only symptoms to be accepted as historical and possibly to be discussed from the psychiatric point of view —the high estimate which Jesus has of himself and perhaps also the hallucination at the baptism—fall far short of proving the existence of mental illness.

[EDITOR's NOTE: The following paragraphs were not included in Schweitzer's doctoral dissertation, but were added when the thesis was printed.]

Emil Rasmussen [53] is much better acquainted with the historical material than the authors previously mentioned and so keeps away from the Fourth Gospel. His desire is to offer a study in comparative psychopathology and to get a glimpse of the mind of Jesus by comparing him with other prophets and men of God. The following are mentioned: the Jewish prophets, Buddha (550?–480 B.C.), Paul (†64?), Mohammed (570?–632), Luther (1483–1546), Sabbatai Zewi of Asia Minor (1626–1676), Swedenborg (1688–1772), the Mahdi (†1885), the saints from the Abruzzi Mountains, Oreste de Amicio (1824–1889), David Lazzaretti, to whom a special study is dedicated,[54] the founders of Bahaism, Mirza Ali Mohammed (†1852) and Mirza Hussein Ali (1817–1892), and the Dane, Sören Kierkegaard (1813–1855).

Rasmussen is of the opinion that these prophets and men of God present a picture of sickness which may be diagnosed by the psychiatrist as precisely the disease of epileptic psychosis.

The Danish author thinks to find in Jesus all the symptoms observed in the old and modern prophets. Jesus is said to show an unparalleled feeling of anxiety, to have gone into a frenzy in the cleansing of the temple, to suffer from hallucinations, to reveal in his contradictory character an unbounded self-consciousness and an abnormal sensual life, to cherish the deluded purpose of suffering for humanity and of being able to atone for it.

---

[53] Emil Rasmussen, *Jesus: A Comparative Study in Psychopathology.* Translated into German by Arthur Rothenburg under the title, *Jesus: Eine vergleichende psychopathologische Studie.* (Leipzig, 1905, 166 pp.).

[54] Emil Rasmussen, *Ein Christus aus unsern Tagen (A Christ from Our Own Time)*, (Leipzig, 1906, 233 pp.).

Since Rasmussen does not discuss the peculiar symptoms of Jesus, but simply subsumes the Nazarene under the prophet-type which he sets up, we can only discuss his diagnosis in general terms. Why he diagnoses it as epilepsy is not quite clear. The author, who is ill-advised from the psychiatric standpoint, identifies the epileptic character—he introduces as paradigms a few great historical figures whose epilepsy has not yet been proved—with isolated psychopathic traits which he thinks he is able to identify in Jesus. Such traits, however, are to be found in all talented people who diverge from the average. Even epileptiform conditions are found in them. The presence of such traits—even assuming that they can be observed in Jesus—are far from justifying a diagnosis of epilepsy.

That Rasmussen himself does not consider his diagnosis to be wholly free from doubt is shown by the remark that he wishes to leave open the possibility that we have to do in Jesus with paranoia.[55] The general psychiatric observations in which he indulges in this connection prove the chaotic condition of the concept of psychiatry in his own mind. The sentence, "The possibility ought not then to be excluded from the outset, that the figures of the prophets and of the Christ might evince, not merely epilepsy, but also paranoia, *dementia paralytica*, and possibly hysteria," seems to make impossible an exhaustive discussion of Rasmussen's psychiatric conception. The medical value of a comparative study of the kind he undertakes is to be rated as exactly zero.

---

[55] Rasmussen, *Jesus: eine vergleichende psychopathologische Studie*, p. 133.

INDEX

Abstinence, 43 f.
Account of a Case of Paranoia (Freud), 16
Alsace, Lower, 20
Alsace, Upper, 19
Amicio, Oreste de, 73
Angels, evil, 48, 66
Angels, good, 48, 66
Anxiety, 73
Apocalypses, 27 f., 49 f., 54, 61, 63, 71
Appearance, public, 47
Atonement, 52
Atropine, 14
Attitude, emotional, 56
Authority, ecclesiastical, 11
Autism, 57

Bahaism, 73
Baptism, 39f., 43 ff., 67
Baruch, 71
Baumann, Julius, 36n.
Bible, 11, 40
Binet-Sanglé, Charles, 12, 14, 36, 42, 43, 44, 45n., 46, 53 f., 54 n., 55–60, 65, 67 ff., 71
Bleuler, 13
Buddha, 73

Caesarea Philippi, 22, 53
Caesar, Julius, 33
Capernaum, 54 n., 68
Captivity, Babylonian, 47
Carpenter, 47, 62
Celsus, 45 n.
Certain Neurotic Mechanisms in Jealousy (Freud), 16
Character of Jesus, The (Julius Baumann), 36n.
Chosen, 63

Christianity, 19, 44 n.
Christianity, ecclesiastical, 26
Christianity, liberal, 25 f.
Christus aus unsern Tagen, Ein (Emil Rasmussen), 73 n.
Concerning the Question of Paranoia (E. Kraepelin), 60 n.
Conclusions of a Psychiatrist (William Hirsch), 36 n.
Constitution, 14
Convictions, Messianic, 19
Copernicus, 11
Council, High, 51
Creation, original, 11
Criticism, higher, 11
Cures, 59, 68

Daimonion, 39
Daniel, Book of, 27, 49 f.
Darwin, 11
David, 25, 41, 47, 50, 61 n., 62 f., 70
David Friedrich Strauss (Ziegler), 35 n., 36
Death, 48
Degenerate, 37, 45 n.
De la Psychose Paranoiaque (J. Lacan), 16
Delirium, chronic, 56 f., 59 n.
Delusions, 12, 14 f., 37, 40–44, 56–59, 63
Dementia, 13
Dementia paralytica, 74
Dementia praecox (Kraepelin's), 57
Democracy, 14
Depression, 38
Derangement, mental, 24
Deuteronomy, 47, 70
Devil, 66 n.
Diagnosis, *à distance*, 14

75

Disciples, 20 ff., 52 f., 64, 66 f., 71
Dissociation, 57
Dogmas, Greek, 26, 45
Dogmatism, 25 f.
Dove, 44

Ecstatic, 11, 24, 43 n.
Elect, 22, 61 f.
Elijah, 21, 53, 66
Emotions, 13, 38
Enoch, 49 f.
Epilepsy, 74
Eschatology, 18, 23 f., 26
Ethics, 26
Étude Critique des États Paranoides (S. Jouannais), 16
Eunuch, 39, 69 f.
Evolution, systematic, 60, 64
Exodus, 66
Extra-Biblical gospels, 45
Ezra, 47

Factors, environmental, 15
Family, 37, 46, 69
Fantasies, 27
Feast, Messianic, 48
Fertility, supernatural, 61, 71
Fig Tree, 38, 41, 70
Folie de Jésus, La (Charles Binet-Sanglé), 36 n.
Freud, 13, 16

Galilee, 22, 48, 67, 71
Gemütsart Jesu, Die (Julius Baumann), 36 n.
Geocentricity, 11
Geschichte der Leben-Jesu-Forschung (Schweitzer), 12, 27
Geschichte Jesus, Die (Noack), 11
Gethsemane, 38, 44, 66
Glory, King of, 50
God, agent of, 59
God, angel of, 44
God, Kingdom of, 48 f., 51 ff., 69 f.
God, servant of, 53
God, Son of, 42 f., 59, 62
God, Spirit of, 45
God, throne of, 48
Gospels, 44 f., 48, 54
Gospels, synoptic, 20
Grandeur, delusions of, 60

Grandeur, ideas of, 15, 24
Grave, 48
Greek Testament, 20
Guggenheim, 20, 22

Haggai, 47
Haiti, 14
Hallucinations, 24, 36 n., 39, 42 ff., 55–60, 65–68, 73
Hallucinations, hypnagogic, 14
Hallucinations, sensory, 12, 14, 39, 42 f., 43 n., 44, 67
Harvard University, 17
Heaven, clouds of, 51
Hebephrenia, 53
Heinroth, 12
Heresy, Schweitzerian, 17
Hippocratic writings, 12
Hiroshima, 18
Hirsch, William, 12, 14, 36, 40–42, 44, 46, 53–56, 58 ff., 62 f., 65
Historical Jesus of Liberal Theology, a Psychotic, The (Hermann Werner), 36 n.
Historische Jesu der liberalen Theologie, ein Geisteskranker, Der (Hermann Werner), 36 n
Hochfelden, 20
Holtzmann, H. J., 27, 35 n., 36
Holtzmann, Oskar, 36 n.
Homosexuality, 13
Hopes, Messianic, 26
Hussein, Mirza, 73
Hybrid, 37, 45 n.
Hysteria, 24, 74

Illusion, 42
Impartiality, 28
Infancy, 45
Injury, ideas of, 15, 64
Insanity of Jesus, The (Charles Binet-Sanglé), 36 n
Instability, emotional, 68
Interpreter, God's, 43
Isaiah, 22, 53, 70
Israel, sins of, 43
Israel, tribes of, 71

James, William, 14, 16
Jeremiah, 17
Jericho, 47

INDEX

Jerusalem, 22, 38, 47, 51 ff., 71
Jesus: A Comparative Study in Psychopathology (Emil Rasmussen), 37 n., 73 n.
Jesus as prophet, 55 n.
Jesus as teacher, 55 n.
Jesus, birth of, 46
Jesus, brothers of, 47
Jesus, childhood of, 46, 53
Jesus Christ from the Standpoint of Psychiatry (George de Loosten), 36 n
Jesus Christus vom Standpunkte des Psychiaters (George de Loosten), 36 n.
Jesus, death of, 51 ff., 65
Jesus, guilt of, 52
Jesus in the Light of Psychiatry (H. Schafer), 56 n.
Jesus in psychiatrischer Beleuchtung (H. Schafer), 56 n.
Jesus, physical appearance of, 47
Jesus, sisters of, 47
Jesus, youth of, 53
Jews, 45
Jews, King of the, 43
John, Gospel of, 38 f., 41 f., 45 f., 48, 54, 65, 73
John the Baptist, 21, 37, 40, 43, 53, 58 f., 61, 64
Jordan, 40, 43, 48, 67
Jouannais, S., 16
Judaism, late, 24, 49, 54, 61 ff., 67, 70 f.
Judas, 51
Judgment, 14, 49, 61, 63
Jülicher, Adolf, 35 n.
Justification, 49

Kaysersberg, 19
Kierkegaard, Sören, 73
Kingdom, Messianic, 18, 20–23, 25, 34, 36 f., 48 f., 52, 61–65, 71
Kingdom of Heaven, 23, 48
Kingdom, supernatural, 23
Kirchhof, H., 16
Klinischer Beitrag zur Differentialdiagnose paranoider Erkrankungen (H. Kirchhof), 16
Kneib, Philipp, 56 n.
Kraepelin, E., 13, 16, 60 n.

Lacan, J., 16
Lang, Wilhelm, 35
Law, 66, 70
Lazzaretti, David, 73
Leben Jesu, Das (David Friedrich Strauss), 35 n.
Leben Jesu fürs deutsche Volk bearbeitet, Das (David Friedrich Strauss), 35 n.
Legend, 66
Leyrer, Pastor, 29
Liberty, 26
Life of Jesus, The (David Friedrich Strauss), 35
Life of Jesus Revised for the German People, The (David Friedrich Strauss), 35 n.
Logos, 45
Lomer, Dr. Georg, 12, 36
Loosten, George de, 12, 14, 36–39, 44, 45 n., 46, 53–56, 58 ff., 62, 65, 68, 71
Love, 25, 26
Luke, Gospel of, 20, 44, 46, 53, 66
Luther, 11, 73

Madelung, 10
Magnan, 56
Mahdi, 73
Malachi, 53
Manic Depressive Insanity and Paranoia (E. Kraepelin), 16
Man, Son of, 27, 38, 41 n., 50–52, 62
Mark, Gospel of, 23, 42, 46–49, 51 ff., 54 n., 66–70
Marriage, 69
Mary, 45 n.
Matthew, Gospel of, 20 f., 23, 38 f., 41 n., 46, 52 f., 66, 71
Medicine, 28
Medicine, Doctor of, 12
Megalomania, 41, 59 f.
Mental Health of Jesus, The (Hermann Werner), 56 n.
Messiah, 14, 21 f., 25, 27, 34 ff., 43 f., 48, 50–54, 55 n., 58 f., 61, 63, 67, 71
Messianic Consciousness of Jesus, The (H. J. Holtzmann), 35 n., 36 n.

Messianic Secret in the Gospels, The (William Wrede), 36 n.
Messianische Bewusstsein Jesu, Das (H. J. Holtzmann), 35 nf.
Messiasgeheimnis in den Evangelien, Das (William Wrede), 36 n.
Ministry, public, 41, 47 f., 64, 68 f.
Miracles, 15
Modern Clinical Psychiatry (A. P. Noyes), 16
Modern Research into the Life of Jesus under the Influence of Psychiatry (Philipp Kneib), 56 n.
Moderne Leben-Jesu-Forschung unter dem Einflusse der Psychiatrie (Philipp Kneib), 56 n.
Mohammed, 33, 73
Money-changers, 39, 65
Morality, 49
Moreira, J., 16
Moses, 66

Napoleon, 33
Nature, transfiguration of, 14
Nazarene, 27, 34 ff., 41, 44, 67, 74
Nazareth, 24, 46, 54 n.
Necromancy, 14
Noack, 11
Noyes, A. P., 16

Paganism, 45 n.
Panthere, 45 n.
Paranoia, 10, 12 f., 15, 24, 27, 40 f., 44, 54 ff., 58 f., 63 f., 74
Paranoia and Homosexuality (Freud), 16
Paranoiafrage, Zur (E. Kraepelin), 60 n.
Paranoia légitime: son origine et nature, La (A. Peixoto and J. Moreira), 16
Paranoia, religious, 42, 43 n., 44 n., 56, 57
Paraphrenia, 13, 60 n.
Park Avenue, 14
Passover, 47 f., 51
Paul, Saint, 24, 26, 47, 67, 73
Peixoto, A., 16
Persecution, 49

Persecution, ideas of, 15, 38, 54 n., 57 f., 60, 64 f., 69
Peter, 67 n., 68
Pfersdorff, Dr., 29
Pharisees, 43, 51, 54 n., 55 n., 65
Pilate, Judgment of, 45 n.
Possession, 69
Posterity, 70
Power, supernatural, 50
Priest, High, 51
Problem of the Last Supper Based upon the Scientific Research of the 19th Century and the Historical Accounts, The (Schweitzer), 22 f.
Prophets, 41, 49 f., 61 f., 73
Prophets, Post-Exilic, 70
Protestant, liberal, 23
Psalms, 67
Psychische Gesundheit Jesu (Hermann Werner), 56 n.
Psychoanalytic Notes upon an Autobiography (Freud), 16
Psychology, 24

Quest of the Historical Jesus, The (Schweitzer), 18, 23

Rabbis, 49
Rasmussen, Emil, 12, 28, 37, 73-74
Reference, ideas of, 24, 41, 58, 59 n.
Religion und Zivilisation vom Standpunkte des Psychiaters (William Hirsch), 36 n.
Repentance, 48, 49
Resurrection, 14, 43, 61
Revelations, 49
Roback, Dr. A. A., 17
Romans, 47

Sabbath, 47, 70
Sacraments, 45
Sacrificial lamb, 43
Saint from the Abruzzi Mountains, 73
Sanhedrin, 22
Savior, 42
Schafer, Dr. H., 56 n.
Schreber, 13
Scribes, 43, 51

Scriptures, 37, 47, 59
Secret of the Messiahship and the Passion. A Sketch of the Life of Jesus, The (Schweitzer), 23
Secretiveness, 44
Séglas, 12
Self-sacrifice, 65
Sermon on the Mount, 26
Servant, suffering, 22
Sex, 37, 39, 69 f.
Siblings, 13
Sidon, 53
Solitude, forty days of, 42
Sources, rabbinical, 61
Spinoza, 11
Spirit, 26
Storm, 70 n.
Strassburg, University of, 20
Strauss, David Freidrich, 27, 34 f., 36 n., 45, 66
Suffering, 53
Supper, Last, 45
Swedenborg, 73
Symphonia Sacra (Widor), 10
Symptoms, catatonic, 57
Symptoms, morbid, 69, 73
Symptoms, psychomotor, 44
Synagogue, 68
Synoptic problem, 20
Systematization, 59

Talmud, 45
Temple, 46, 53, 65, 73
Temptations, 66
Testament, New, 18, 23
Testament, Old, 37, 59 n., 66 f., 69
Theology, 18, 21
Theology, historical, 28
Theomanic possession, 44

Thomas, Saint, 17
Torment, eternal, 48
Tractatus Theologico-Politicus (Spinoza), 11
Tradition, 67
Transfiguration, 42, 61, 66
Tribulation, days of, 22
Truth, 24, 25
Tyre, 53

Understanding Greek, 45
Utopia, 38

Varieties of Religious Experience, (William James), 16
Vergleichende psychopathologische Studie, Eine (Arthur Rothenburg), 37 n., 73 n., 74 n.

War Jesu Ekstatiker? (Oskar Holtzmann), 36 n.
Was Jesus an Ecstatic? (Oskar Holtzmann), 36 n.
Way, prophetic, 49
Weltanschauung, 19
Werner, Hermann, 36 n., 56 n.
Widener Library, 17
Widor, Charles Marie, 10
Wilderness, 40, 42, 44, 59 f., 66, 68
Wisdom of Solomon, 70
Witness, God's, 43
Woes of the Messiah, 49, 64
Wollenberg, Dr., 29
Wrede, William, 36 n.

Zachariah, 47
Zerubbabel, 47
Zewi, Sabbatai, 73
Ziegler, Theobald, 35 n., 36 n.